Compendium of Plant Genomes

Series editor

Prof. Chittaranjan Kole
Raja Ramanna Fellow, Department of Atomic Energy, Government of India, Kalyani, India

Whole-genome sequencing is at the cutting edge of life sciences in the new millennium. Since the first genome sequencing of the model plant Arabidopsis thaliana in 2000, whole genomes of about 70 plant species have been sequenced and genome sequences of several other plants are in the pipeline. Research publications on these genome initiatives are scattered on dedicated web sites and in journals with all too brief descriptions. The individual volumes elucidate the background history of the national and international genome initiatives; public and private partners involved; strategies and genomic resources and tools utilized; enumeration on the sequences and their assembly; repetitive sequences; gene annotation and genome duplication. In addition, synteny with other sequences, comparison of gene families and most importantly potential of the genome sequence information for gene pool characterization and genetic improvement of crop plants are described. Interested in editing a volume on a crop or model plant? Please contact Dr. Kole, Series Editor, at ckole2012@gmail.com.

More information about this series at http://www.springer.com/series/11805

Manoj Prasad
Editor

The Foxtail Millet Genome

Springer

Editor
Manoj Prasad
Plant Molecular Biology
National Institute of Plant Genome Research
New Delhi, Delhi
India

ISSN 2199-4781 ISSN 2199-479X (electronic)
Compendium of Plant Genomes
ISBN 978-3-319-65616-8 ISBN 978-3-319-65617-5 (eBook)
DOI 10.1007/978-3-319-65617-5

Library of Congress Control Number: 2017950248

Printed on acid-free paper

This Springer imprint is published by Springer Nature
The registered company is Springer International Publishing AG
The registered company address is: Gewerbestrasse 11, 6330 Cham, Switzerland

This book series is dedicated to
my wife Phullara, and our children Sourav,
and Devleena

Chittaranjan Kole

Preface to the Series

Genome sequencing has emerged as the leading discipline in the plant sciences coinciding with the start of the new century. For much of the twentieth century, plant geneticists were only successful in delineating putative chromosomal location, function, and changes in genes indirectly through the use of a number of "markers" physically linked to them. These included visible or morphological, cytological, protein, and molecular or DNA markers. Among them, the first DNA marker, the RFLPs, introduced a revolutionary change in plant genetics and breeding in the mid-1980s, mainly because of their infinite number and thus potential to cover maximum chromosomal regions, phenotypic neutrality, absence of epistasis, and codominant nature. An array of other hybridization-based markers PCR-based markers, and markers based on both facilitated construction of genetic linkage maps, mapping of genes controlling simply inherited traits and even gene clusters (QTLs) controlling polygenic traits in a large number of model and crop plants. During this period a number of new mapping populations beyond F_2 were utilized and a number of computer programs were developed for map construction, mapping of genes, and for mapping of polygenic clusters or QTLs. Molecular markers were also used in studies of evolution and phylogenetic relationship, genetic diversity, DNA-fingerprinting, and map-based cloning. Markers tightly linked to the genes were used in crop improvement employing the so-called marker-assisted selection. These strategies of molecular genetic mapping and molecular breeding made a spectacular impact during the last one and a half decades of the twentieth century. But still they remained "indirect" approaches for elucidation and utilization of plant genomes since much of the chromosomes remained unknown and the complete chemical depiction of them was yet to be unraveled.

Physical mapping of genomes was the obvious consequence that facilitated development of the "genomic resources" including BAC and YAC libraries to develop physical maps in some plant genomes. Subsequently, integrated genetic-physical maps were also developed in many plants. This led to the concept of structural genomics. Later on, emphasis was laid on EST and transcriptome analysis to decipher the function of the active gene sequences leading to another concept defined as functional genomics. The advent of techniques of bacteriophage gene and DNA sequencing in the 1970s was extended to facilitate sequencing of these genomic resources in the last decade of the twentieth century.

As expected, sequencing of chromosomal regions would have led to too much data to store, characterize, and utilize with the-then available computer software could handle. But development of information technology made the life of biologists easier by leading to a swift and sweet marriage of biology and informatics and a new subject was born—bioinformatics.

Thus, evolution of the concepts, strategies, and tools of sequencing and bioinformatics reinforced the subject of genomics—structural and functional. Today, genome sequencing has traveled much beyond biology and involves biophysics, biochemistry, and bioinformatics!

Thanks to the efforts of both public and private agencies, genome sequencing strategies are evolving very fast, leading to cheaper, quicker and automated techniques right from clone-by-clone and whole-genome shotgun approaches to a succession of second-generation sequencing methods. Development of software of different generations facilitated this genome sequencing. At the same time, newer concepts and strategies were emerging to handle sequencing of the complex genomes, particularly the polyploids.

It became a reality to chemically—and so directly—define plant genomes, popularly called whole-genome sequencing or simply genome sequencing.

The history of plant genome sequencing will always cite the sequencing of the genome of the model plant *Arabidopsis thaliana* in 2000 that was followed by sequencing the genome of the crop and model plant rice in 2002. Since then, the number of sequenced genomes of higher plants has been increasing exponentially, mainly due to the development of cheaper and quicker genomic techniques and, most importantly, development of collaborative platforms such as national and international consortia involving partners from public and/or private agencies.

As I write this preface for the first volume of the new series "Compendium of Plant Genomes", a net search tells me that complete or nearly complete whole-genome sequencing of 45 crop plants, eight crop and model plants, eight model plants, 15 crop progenitors and relatives, and three basal plants are accomplished, the majority of which are in the public domain. This means that we nowadays know many of our model and crop plants chemically, i.e., directly, and we may depict them and utilize them precisely better than ever. Genome sequencing has covered all groups of crop plants. Hence, information on the precise depiction of plant genomes and the scope of their utilization is growing rapidly every day. However, the information is scattered in research articles and review papers in journals and dedicated web pages of the consortia and databases. There is no compilation of plant genomes and the opportunity of using the information in sequence-assisted breeding or further genomic studies. This is the underlying rationale for starting this book series, with each volume dedicated to a particular plant.

Plant genome science has emerged as an important subject in academia, and the present compendium of plant genomes will be highly useful both to students and teaching faculties. Most importantly, research scientists involved in genomics research will have access to systematic deliberations on the plant genomes of their interest. Elucidation of plant genomes is not only of interest for the geneticists and breeders, but also for practitioners of an array of plant science disciplines, such as taxonomy, evolution, cytology, physiology, pathology, entomology, nematology, crop production, biochemistry, and obviously bioinformatics. It must be mentioned that information regarding each plant genome is ever-growing. The contents of the volumes of this compendium are therefore focusing on the basic aspects of the genomes and their utility. They include information on the academic and/ or economic importance of the plants, description of their genomes from a molecular genetic and cytogenetic point of view, and the genomic resources developed. Detailed deliberations focus on the background history of the national and international genome initiatives, public and private partners involved, strategies and genomic resources and tools utilized, enumeration on the sequences and their assembly, repetitive sequences, gene annotation, and genome duplication. In addition, synteny with other sequences, comparison of gene families, and, most importantly, potential of the genome sequence information for gene pool characterization through genotyping by sequencing (GBS) and genetic improvement of crop plants have been described. As expected, there is a lot of variation of these topics in the volumes based on the information available on the crop, model, or reference plants.

I must confess that as the series editor it has been a daunting task for me to work on such a huge and broad knowledge base that spans so many diverse plant species. However, pioneering scientists with life-time experience and expertise on the particular crops did excellent jobs editing the respective volumes. I myself have been a small science worker on plant genomes since the mid-1980s and that provided me the opportunity to personally know several stalwarts of plant genomics from all over the globe. Most, if not all, of the volume editors are my longtime friends and colleagues. It has been highly comfortable and enriching for me to work with them on this book series. To be honest, while working on this series I have been and will remain a student first, a science worker second, and a series editor last. And I must express my gratitude to the volume editors and the chapter authors for providing me the opportunity to work with them on this compendium.

I also wish to mention here my thanks and gratitude to the Springer staff, Dr. Christina Eckey and Dr. Jutta Lindenborn in particular, for all their constant and cordial support right from the inception of the idea.

I always had to set aside additional hours to edit books besides my professional and personal commitments—hours I could and should have given to my wife, Phullara, and our kids, Sourav, and Devleena. I must mention that they not only allowed me the freedom to take away those hours from them but also offered their support in the editing job itself. I am really not sure whether my dedication of this compendium to them will suffice to do justice to their sacrifices for the interest of science and the science community.

Kalyani, India Chittaranjan Kole

Preface

Foxtail millet [*Setaria italica* (L.) P. Beauv.] is one of the world's oldest domesticated crops, and it has become a major crop adapted to tropical, sub-tropical, and temperate regions of India, China, and other parts of Asia, North Africa, and the Americas. It has a small diploid genome ($\sim$515 Mb), short growing cycle 0–80 days), self-fertilization, and small morphological stature. Its prolific seed production per plant has made this crop a novel model for grass research. Belonging to the tribe Paniceae in the subfamily Panicoideae, foxtail millet is evolutionarily close to major cereals including sorghum (*Sorghum bicolor*), maize (*Zea mays*), rice (*Oryza sativa*), and candidate biofuel crops such as switchgrass (*Panicum virgatum*), napiergrass (*Pennisetum purpureum*), and pearl millet (*Pennisetum glaucum*). These biofuel crops possess polyploid genomes, long generation times, and large statures, and hence are difficult to study. Therefore, foxtail millet serves as an excellent surrogate genome for studying the genetics and genomics of bioenergy grasses. Efficient biomass productivity and improved water- and nitrogen-use efficiencies are attributable to C_4 photosynthesis, which is assumed to be a prime constituent of exceptional productivity among cereals including maize and sorghum, and bioenergy grasses such as pearl millet and switchgrass. C_4 photosynthesis has evolved as a result of 66 independent plant lineages, and therefore their study is important for understanding the biochemistry, physiology, and molecular biology of C_4 photosynthesis, which could be useful in transferring this system to C_3 species.

Extensive research has been carried out on maize and sugarcane to elucidate the C_4 biology; however, the genome complexity of these crops has resulted in little success. With the availability of reference genomes, haplotype maps, and small genome size, foxtail millet is accentuated as a model system for molecular characterization of C_4 genetics and physiology. In addition, being a staple crop of arid and semi-arid regions of the world, foxtail millet is well known for its tolerance to several abiotic stresses, particularly drought and salinity. This characteristic feature is more important in studying the genetics and genomics of stress tolerance and, therefore, foxtail millet is also considered as a model crop for abiotic stress biology. Recently, the health-benefiting properties of millets have gained importance in nutritional research and, importantly, grains of foxtail millet are reported to be rich in protein, dietary fiber, and energy content, which are three- to fivefold higher than major cereals such as rice, wheat, and maize. Altogether, foxtail millet holds the thrust for establishing climate-resilient agriculture in order to serve nutrient-rich food and fodder to the ever-growing global population.

India tops the list in global millet production; however, pearl millet and finger millet are the top two varieties extensively cultivated in India. Despite the prominent attributes encompassed by foxtail millet, breeding technologies used in this crop are far behind those of pearl millet and finger millet, and other major cereals such as rice and wheat. Foxtail millet has also received very little research attention worldwide, and less effort has been invested towards dissecting the genetic determinants of the prominent traits which are important for improvement of this model species as well as other millets, cereals, and biofuel crops. In this context, the book enumerates the national and international efforts invested in delineating structural, functional, and nutritional genomics of this important crop. The book has 11 chapters describing the general introduction to foxtail millet followed by sequencing efforts, and structural and functional genomics. Chapter 1 provides a general introduction to the crop by outlining its agro-economic importance, origin, distribution, taxonomy, and cytology. Chapter 2 summarizes the sequencing efforts, the outcomes, and the application of sequence information in expediting genomics studies in foxtail millet. Chapter 3 provides interesting information on the impact of transposable elements on distribution pattern and evolution of foxtail millet and its wild progenitor, green foxtail. Chapter 4 summarizes the genetic and genomic resources available in this crop for use in the improvement of foxtail millet per se as well as other related millets, biofuel crops, and cereals. Chapter 5 describes the breeding strategies implemented in foxtail millet and gives an elaborate picture of how the same is being performed in India. Chapters 6 and 7 discuss the association mapping studies and genetic structure analysis of foxtail millet performed using high-throughput molecular markers. Following this, Chap. 8 covers the genetic determinants of abiotic stress tolerance, Chap. 9 the genetic transformation system available in both foxtail millet and green foxtail, Chap. 10 the nutritional potential of this crop and the relevance of nutritional genomics in delineating the health benefiting traits, and Chap. 11 the small RNA biology of foxtail millet. Altogether, the book serves as a primary resource material for researchers, breeders, and students working on millet genomics.

I personally thank Prof. Chittaranjan Kole for giving me the opportunity to edit this interesting book. I also acknowledge the help extended by one of my research scholars, Dr. Muthamilarasan, for his expert assistance in editing and finalizing the chapters.

New Delhi, India Manoj Prasad

Contents

Editor and Contributors

About the Editor

Dr. Manoj Prasad (DOB: 16 December 1970) is working as a Senior Scientist at National Institute of Plant Genome Research (NIPGR), New Delhi, India. He received his Ph.D. degree from Department of Botany, University of Calcutta in 1999. Prior to working at NIPGR, Dr. Prasad served as Alexander von Humboldt Fellow at IPK, Gatersleben, Germany (2000–2003). His research interests include Plant Molecular Genetics and Genomics, and Stress Biology. Dr. Prasad has published more than 120 research papers, reviews, and book chapters in reputed journals. He is serving as Associate Editor in renowned international journals of high repute. He is also a Fellow of three National Academies (INSA, New Delhi, NASI, Allahabad, and NAAS, New Delhi) and has received several awards and honors, including National Bioscience Award of Department of Biotechnology, Government of India, Hiralal Chakraborty Memorial Award, Indian Science Congress Association, India, Alexander von Humboldt Fellow, Germany, and Young Scientist Award and Associateship of NAAS, New Delhi, India.

Contributors

Tirthankar Bandyopadhyay National Institute of Plant Genome Research (NIPGR), New Delhi, India

Charu Lata CSIR-National Botanical Research Institute, Lucknow, India; Academy of Scientific and Innovative Research (AcSIR), New Delhi, India

Kenji Fukunaga Prefectural University of Hiroshima, Shobara, Japan

Gunaseelen Hari-Gowthem National Institute of Plant Genome Research (NIPGR), New Delhi, India

K. Hariprasanna ICAR-Indian Institute of Millets Research, Hyderabad, India

Jinu Jacob ICAR-Indian Institute of Millets Research, Hyderabad, India

Vandana Jaiswal Laboratory of Translational and Evolutionary Genomics, School of Life Science, Jawaharlal Nehru University, New Delhi, India

Mehanathan Muthamilarasan National Institute of Plant Genome Research (NIPGR), New Delhi, India

Parashuram Patroti ICAR-Indian Institute of Millets Research, Hyderabad, India

Manoj Prasad National Institute of Plant Genome Research (NIPGR), New Delhi, India

Radha Shivhare CSIR-National Botanical Research Institute, Rana Pratap Marg, Lucknow 226001, India

Shweta Shweta National Institute of Plant Genome Research (NIPGR), New Delhi, India

Roshan Kumar Singh National Institute of Plant Genome Research (NIPGR), New Delhi, India

Priyanka Sood National Institute of Plant Genome Research (NIPGR), New Delhi, India

K. B. R. S. Visarada ICAR-Indian Institute of Millets Research, Hyderabad, India

Amita Yadav National Institute of Plant Genome Research (NIPGR), New Delhi, India

Chandra Bhan Yadav National Institute of Plant Genome Research (NIPGR), New Delhi, India

Foxtail Millet: An Introduction

1

Roshan Kumar Singh, Mehanathan Muthamilarasan
and Manoj Prasad

Abstract

Foxtail millet (*Setaria italica* L.) is a versatile crop known for being genetically closely related to biofuel grasses, for its C_4 photosynthesis, and for its tolerance to abiotic stresses. These attributes have made this crop a model system and, in view of this, the genome of foxtail millet has been sequenced. Among millets, foxtail millet is the only crop possessing rich genetic and genomic resources, and globally it is the second most cultivated millet next to pearl millet. In the context of its importance in agronomic and research terms, the present chapter summarizes the origin, domestication, phylogeny, and agroeconomic importance of foxtail millet.

1.1 Introduction

The term 'millets' refers to a diverse group of annual cereal crops that characteristically produce small seeds. They include several food, fodder, and biofuel grasses, such as foxtail millet (*Setaria italica*), finger millet (*Elucine coracana*), pearl millet (*Pennisetum glaucum*), proso millet (*Panicum miliaceum*), kodo millet (*Paspalum scorbiculatum*), barnyard millet (*Echinocloa* sp.), etc. (Dwivedi et al. 2012). The major distinctive feature of the millets is their adaptability to cope with adverse agro-ecological conditions such as a semi-dry environment and nutritionally poor soil, the requirement of minimal inputs, and highly nutritious seed content (Lata et al. 2013). Millets require very little water for their cultivation—just around 25–30% of the annual rainfall required by crops such as rice and sugarcane. Thus, millets do not require irrigation and power for their production. In addition, millets also not require any synthetic fertilizers and are completely pest-free crop as none of the millets attracts any pests. Thus, the production of millets is very economical for farmers because of almost nil expenditure on irrigation, fertilizers, and pesticides. Importantly, seeds of most millets can be stored for longer period and are not affected by storage pests. Nutritionally, millets are several times superior to other cereal crops such as rice and wheat. They are rich in minerals

R.K. Singh · M. Muthamilarasan · M. Prasad (✉)
National Institute of Plant Genome Research (NIPGR), Aruna Asaf Ali Marg, New Delhi 110067, India
e-mail: manoj_prasad@nipgr.ac.in

M. Prasad (ed.), *The Foxtail Millet Genome*, Compendium of Plant Genomes,
DOI 10.1007/978-3-319-65617-5_1

including iron, calcium, potassium, zinc, and magnesium, vitamins, crude edible fibers, and are gluten-free with low glycemic index (GI). Millet consumption enables a reduced rate of glucose release over a longer duration of time and thus, because of the low GI, their routine ingestion decreases the risk of diabetes mellitus. Regardless of their excellent qualities, millet consumption as food is limited to conventional consumers, particularly rural populations, only in certain parts of the world. This is primarily because of the lack of awareness of their nutritious qualities among most people and the non-availability of customer friendly and ready-made millet-based products. In the past few years, global attention and efforts have been applied to millets to acquire expedient and value-added processed products in consumer markets.

Foxtail millet (*S. italica*) is an important crop used as a staple food in many parts of the world including arid and semi-arid areas of China, some part of India and Japan, and is grown for silage and hay in South and North America. Foxtail millet is commonly known in India as Kangni (Hindi), Kang (Gujrati), Navane (Kannada), Kaon dana (Bengali), Kavalai, and Tenai (Tamil), Kangam (Oriya); as Su, Xiaomi, Shao-mi, Kou wei tsao in China; Awa in Japan; Siberian millet and Dawa in Indonesia; Mohar in Russia; Millet des oiseaux and millet d'ItalieIt in France; Panico, Milho panico, and Milho panico de Itálica in Portugal; and Kimanga in Kenya. It is the second most cultivated millet after pearl millet. Worldwide, total millet production amounts to 29.8 million tonnes, the contribution of India being highest at 10.3 million tonnes (FAOSTAT data 2012; http://faostat.fao.org/). Because of its economic value the Joint Genome Institute (JGI) of the Department of Energy, USA and BGI (formerly Beijing Genome Initiative), China independently sequenced its genome, and the draft genome sequence was released in 2012. The most likely gene number for foxtail millet is around 24,000–29,000, which is in line with gene complements of other diploid grasses such as rice and sorghum (Bennetzen et al. 2012; Zhang et al. 2012). It has been recently identified as an excellent model crop for the study of genetic and molecular aspect of abiotic stress tolerance mechanism and physiology of C_4 photosynthesis process because of its small diploid genome ($\sim$515 Mb; $2n = 2x = 18$), self-pollinating, short growing cycle (50–80 days), small plant architecture and prolific seed production per plant, low repetitive DNA content (30%), and a highly conserved genome structure relative to the ancestral grass lineage such as switchgrass (*Panicum virgatum*), napiergrass (*Pennisetum purpureum*), and pearl millet (*P. glaucum*) (Sivaraman and Ranjekar 1984; Devos et al. 1998; Li and Brutnell 2011; Lata et al. 2013).

1.2 Taxonomy and Morphological Description

The genus *Setaria* has approximately 125 species widely distributed in warm and temperate parts of the world, and this includes *S. italica* (foxtail millet). This genus belongs to the subfamily Panicoideae and the tribe Paniceae. It contains grain, wild, and weed species with different breeding systems, life cycles, and ploidy levels (Lata et al. 2013). The genome of *S. italica* and *S. viridis* (green foxtail) is designated as AA genome with $2n = 2x = 18$ (Benabdelmouna et al. 2001a). Weedy tetraploid species *Setaria faberii* and *Setaria verticillata* have AABB genome, probably originating from a natural cross between *S. viridis* and another diploid species, *Setaria adhaerans* (Benabdelmouna et al. 2001a, b). *Setaria grisebachii* from Mexico has been identified as CC genome diploid species (Wang et al. 2009). *Setaria queenslandica* is the only autotetraploid (AAAA genome) species in genus *Setaria* whereas other polyploid species such as *Setaria pumila* and *Setaria pallide-fusca* do not contain the AA genome (Benabdelmouna et al. 2001a, b; Benabdelmouna and Darmency 2003). The taxonomic hierarchy of foxtail millet is as follows:

Kingdom: Plantae
Subkingdom: Tracheobionta

Superdivision: Spermatophyta
Division: Magnoliophyta
Class: Liliopsida
Subclass: Commelinidae
Order: Cyperales
Family: Poaceae
Genus: *Setaria*
Species: *italica*

Foxtail millet has a typical domesticated plant architectural form consisting of a single stalk or a few tillers, with large inflorescences that mature more or less at the same time. A fully-grown foxtail millet plant measures around 120–200 cm (3.9–6.6 ft) in height with slim, erect, and leafy stems (Fig. 1.1). The smooth and hairless leaves are arc-broad, whereas culms are erect and slender with hollow internodes. The stems are topped by a bristly panicle which is long (5–30 cm long) and mostly reddish or purplish (Fig. 1.1). They give the panicle the appearance of a fox's tail, which is the common name for cultivated millets belonging to the genus *Setaria*. The inflorescence is a contracted panicle, often nodding at the top; on account of its short branches, it resembles a spike. The spikelets are crowded and mixed with stiff bristles, the latter representing branches on which no spikelets are developed. Each spikelet contains only one flower with a yellow pistil. It has a short generation time (depending on the sample, approximately 5–8 weeks from planting to flowering, 8–15 weeks from planting to seed maturity) and can produce hundreds of seeds per inflorescence (Reddy et al. 2006).

A single inflorescence can produce hundreds of small convex seeds measuring about 2 mm in diameter, encased in a thin, papery hull which is easily removed in threshing. The color of the seeds varies greatly between varieties. The non-dormant seeds germinate readily in a glasshouse at densities up to 100 plants per square

Fig. 1.1 Foxtail millet plants cultivated in field condition. **a** Fully grown foxtail millet plants with inflorescence. **b** Bristle panicle. **c** Mature seeds

meter or in field conditions in temperate or tropical regions (Dekker 2003). It is a summer crop, typically planted in late spring. Harvesting for hay or silage can be carried out after 65–70 days (typical yield is 15,000–20,000 kg/ha of green matter or 3,000–4,000 kg/ha of hay), and for grain after 75–90 days (typical yield is 800–900 kg/ha of grain). Early maturity and excellent water use efficiency (WUE) make it suitable for cultivation in dry and arid regions.

1.3 Origin and Distribution

Foxtail millet is one of the oldest cultivated crops in the world, the earliest archaeo-botanical macro remains indicating its origin in Cishan and Peiligang ruins in Yellow River Valley in the northern province of China, approximately 7,400–7,900 years before present (BP) (Doust et al. 2009). Green foxtail (*S. viridis*) is the wild ancestral form of modern cultivated foxtail millet (*S. italica*) (Wang et al. 1995; Le Thierry d'Ennequin et al. 2000). A combination of foxtail millet and proso millet (*P. miliaceum*) cultivation was practised by the people in ancient China (Lu et al. 2009; Yang et al. 2012; Lata et al. 2013). It has been proposed by Vavilov (1926) that the prime center of evolution and diversification of foxtail millet was East Asia, specifically China and Japan (Fig. 1.2). Another school of thought hypothesized that foxtail millet was cultivated independently in arid and drier part of Europe and Middle East Asia approximately 4000 years BC as indicated by archaeological remains, ribosomal DNA, and isozyme and phenotypic variance (Jusuf and Pernes 1985; Hunt et al. 2008; Austin 2006). Neither cultivated nor wild samples of foxtail millet showed a clear differentiation of population structure, but both samples from China were the most genetically diverse, which supports the idea of the monophyletic origin of foxtail millet in China (Le Thierry d'Ennequin et al. 2000). Tillering and panicle shape were associated with domestication as indicated by quantitative trait loci (QTL mapping of candidate genes, whereas the origin of waxy phenotype in foxtail millet was associated with human selection (Doust et al. 2005).

Currently, foxtail millet is distributed in most of China, some parts of India, USA, Canada, the Korean Peninsula, Japan, Indonesia, Australia, and the northern part of Africa (Doust et al. 2009; Li and Brutnell 2011). In the United States, foxtail millet is primarily produced in the northern and western Great Plains, midwest,

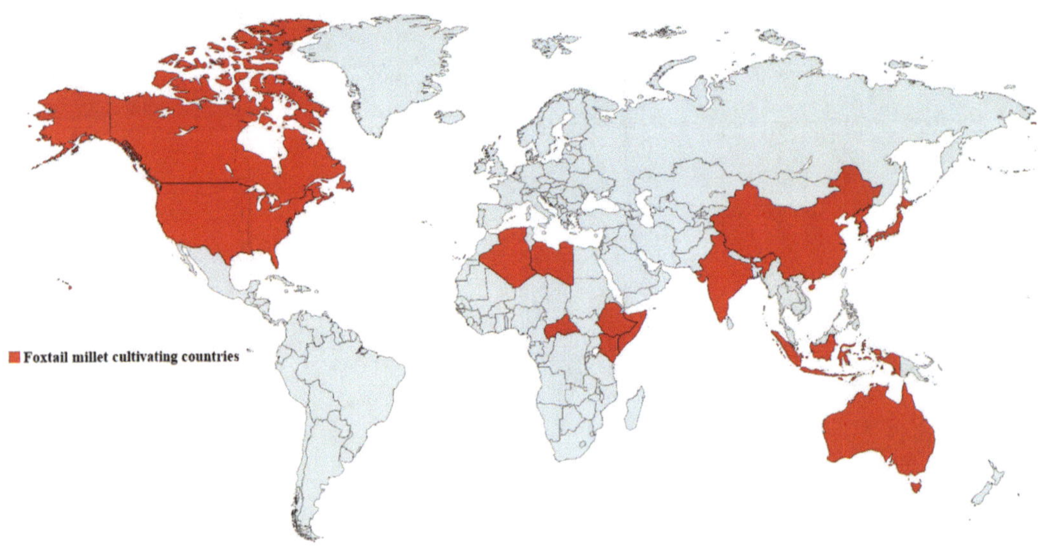

Fig. 1.2 Geographical distribution of foxtail millet cultivation. World map representing the regions highlighted in red where the foxtail millet cultivation is in practice

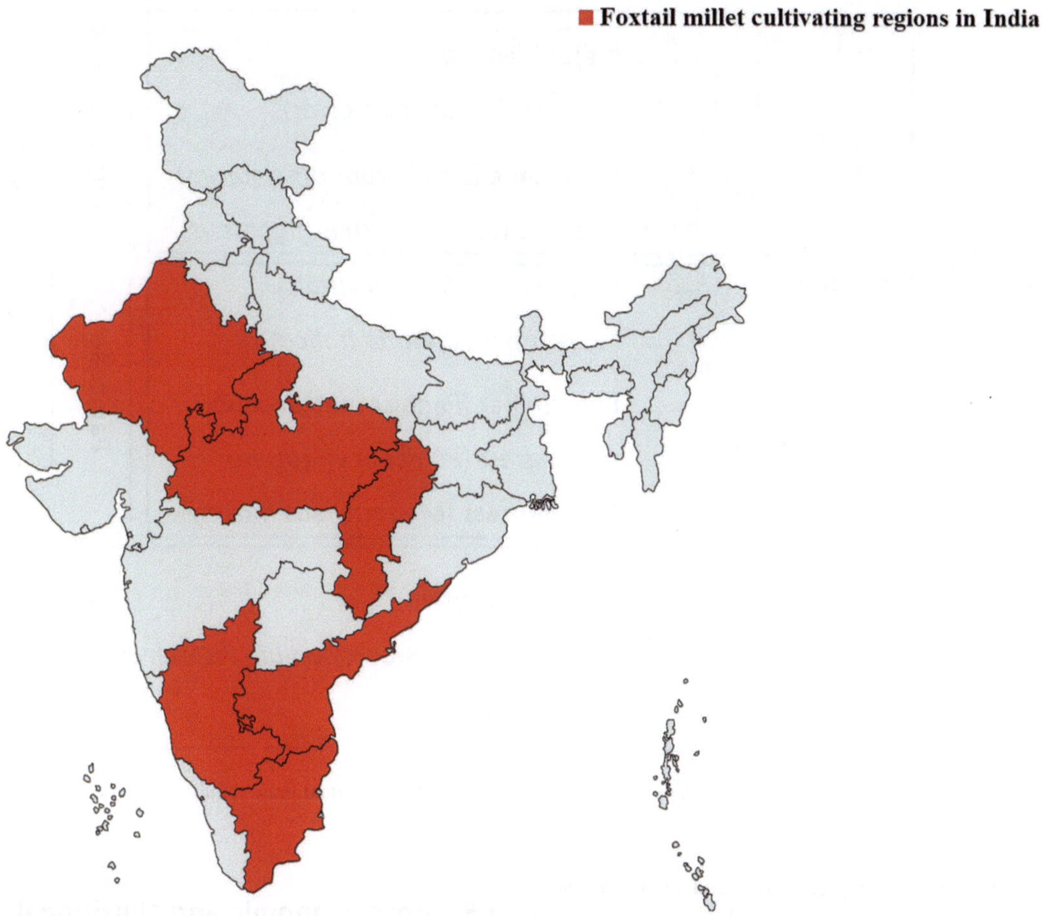

Fig. 1.3 Geographical distribution of foxtail millet cultivation in India. Map of India representing the regions highlighted in red where foxtail millet cultivation is practiced

Colorado, the Dakotas, Kansas, Wyoming, and Nebraska (Oelke et al. 1990; Baltensperger 1996). Foxtail millet can be found grown mainly for feed purposes in some part of southern Europe. In India, it is cultivated primarily in Karnataka, Andhra Pradesh, Rajasthan, Madhya Pradesh and Chhattisgarh, and Tamil Nadu (Fig. 1.3).

1.4 Phylogeny

An inclusive phylogenetic study based on nuclear as well as organellar DNA study revealed foxtail millet to be very closely related to green foxtail (Giussani et al. 2001; Doust et al. 2009). This also supports the hypothesis that, because of the course of evolution and domestication selection, foxtail millet has evolved from green foxtail (Wang et al. 1995; Le Thierry d'Ennequin et al. 2000; Doust et al. 2009). Both the species of genus *Setaria* are part of a larger monophyletic clade consisting of approximately 300 species, all with identical inflorescence (spikelets and bristles) features (Zuloaga et al. 2000). Foxtail millet morphologically differs from its wild ancestor green foxtail in its enlarged bristles with complex branching pattern, flowering synchrony, condensed axillary and basal vegetative branching, and loss of seed dormancy and disarticulation (Doust et al. 2004; Doebley et al. 2006; Lata et al. 2013). Furthermore, it is also closely related to several other biofuel crops such

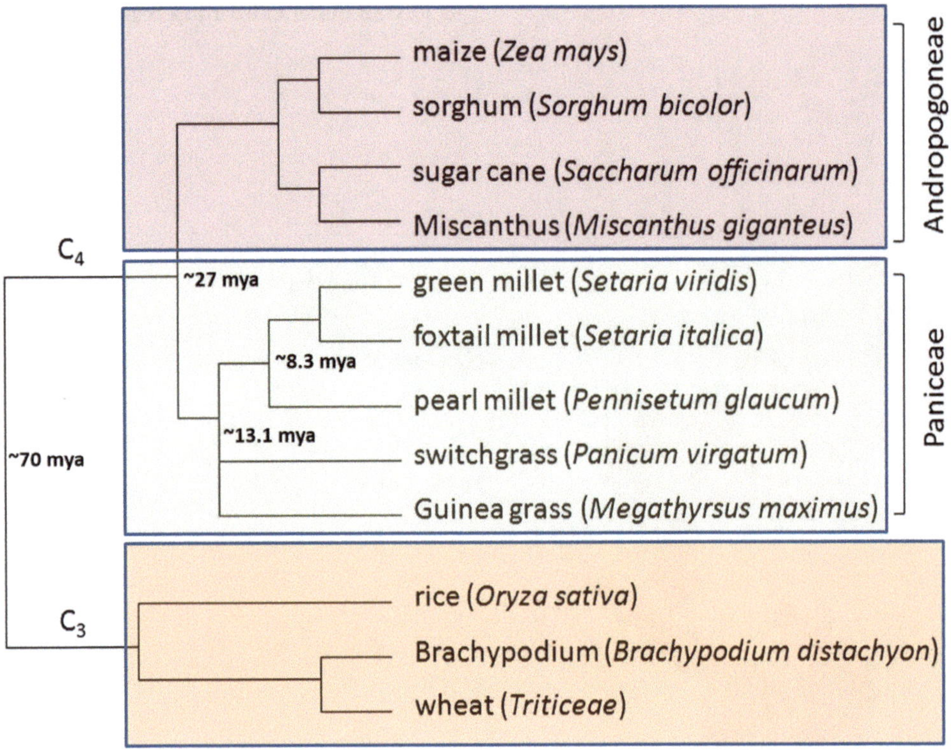

Fig. 1.4 Phylogenetic relationship of *Setaria italica* relative to selected important grass family members

as switchgrass, napier grass, and pearl millet at the genome level, and hence has been suggested to represent a relevant model for this class of crops (Doust et al. 2009; Lata et al. 2013). *Setaria* last shared a common ancestor with pearl millet ~8.3 million years ago and with switchgrass and proso millet ~13.1 million years ago (Vicentini et al. 2008). Later, it was separated from sorghum and maize ~27 million years ago. It was estimated that a whole genome duplication event had occurred to all members of the grass family ~70 million years ago, before the separation of *Setaria* from maize and sorghum (Fig. 1.4). There is highly conserved collinearity between genomic regions of *S. italica* and rice (71.8%), maize (86.7%), *Brachypodium* (61.5%), and sorghum (72.1%), which indicates some close evolutionary relationships between these grasses (Zhang et al. 2012).

1.5 Agro-Economic and Nutritional Importance

Being a C_4 photosynthetic crop, foxtail millet is naturally equipped with excellent WUE and nitrogen use efficiency, and, in addition, several morpho-physiological traits including dense and deep root systems, smaller leaf area, and thickening of cell walls which were thought to lead to durable tolerance to a range of abiotic stresses—mainly drought, heat, and salinity (Lata et al. 2013; Diao et al. 2014). In addition to that, the grains of foxtail millet require only 26% of their grain weight in water to germinate, whereas other major cereals such as rice, wheat, and maize require a minimum of 45% of their grain weight. Similarly, to produce 1 g dry biomass, foxtail millet requires only 257 g of water, which is the minimum among other cereals, as wheat and maize requires 470 and 510 g, respectively (Diao et al. 2014). These properties of foxtail millet demonstrate the important agronomical features and climate resilient characteristics.

Further investigation of genetics and genomics of abiotic stress mechanisms had revealed that foxtail millet has a novel gene as well as known stress-responsive genes which participate in stress tolerance (Muthamilarasan and Prasad 2015). Among abiotic stresses, heat, drought and salt stress are the instant impact of climate change and, in this regard, the research community is actively involved in understanding the genetics and genomics of crops tolerant to these stresses, with the goal of engineering these traits in stress-susceptible plants (Kole et al. 2015). Understanding the significance of foxtail millet as a model crop for abiotic stress biology, efforts have made to investigate the role(s) of several stress-responsive genes including *AP2/ERF* (Lata et al. 2011, 2014), *NAC* (Puranik et al. 2011, 2013), *WD40* (Mishra et al. 2012, 2014), C_2H_2 *zinc finger* (Muthamilarasan et al. 2014a), *MYB* (Muthamilarasan et al. 2014b), *DCL*, *AGO*, and *RDR* (Yadav et al. 2015), *WRKY* (Muthamilarasan et al. 2015), and *ADP-ribosylation factors* (Muthamilarasan et al. 2016b), *HSP* (Singh et al. 2016), and *SET* (Yadav et al. 2016) in response to several abiotic stresses.

Foxtail millet grain is rich in protein (14–16%), crude fat (6–8%), and iron along with zinc and calcium (Muthamilarasan and Prasad 2015; Muthamilarasan et al. 2016a). Not only is the biological value of digestible protein higher than rice and wheat; seven of the eight essential amino acids, which cannot be synthesized by the human body, are higher in foxtail millet (Zhang et al. 2007). A grain of foxtail millet contains approximately 2.5 times the edible fiber found in rice, and is thus a promising source of edible fiber, which is important for intestine and stomach health (Liang et al. 2010). Foxtail millet bran contains 8–10% crude oil and is rich in linoleic (66.5%) and oleic (13.0%) acids (Liang et al. 2010). Through the long cultivation and utilization as food, were fashioned using foxtail millet seeds. In China its flour is used to make bread, chapattis, pancakes, and snacks. Steamed bread made from composite flour some different methods of consumption containing foxtail millet, wheat, and soybean has gained prominence in Northern China (Diao et al. 2014). In India, its flour is mainly used to make bread, chapattis, cookies, and snacks. However, in many countries it is still grown generally for fodder, birdseed, silage, and hay.

Recently, the health-benefiting properties of millets have gained importance in nutritional and medicinal research. Foxtail millet is widely used not only as an energy source for pregnant and lactating women, but also for sick people and children, and especially for people with diabetes. It is reported to reduce the blood sugar concentration in women diabetics (Sema and Sarita 2002; Dwivedi et al. 2012). Foxtail millet consumption is very important for type-II diabetes patients as it helps in reducing blood glucose concentration, glycosylated haemoglobin, and serum lipids (Thathola et al. 2010). In China it is used to cure rheumatism. The germinated seed of yellow-seeded cultivars is astringent, digestive, emollient, and stomachic. It is used in the treatment of dyspepsia, poor digestion, and food stagnancy in the abdomen (www.agrisources.com/herbs/setariaitalica). Thus, foxtail millet is a versatile crop.

Acknowledgements Studies on millet genomics in Dr. Manoj Prasad's laboratory are supported by Science and Engineering Research Board (SERB), Department of Science and Technology (DST), Govt. of India [Grant No. EMR/2015/000464], by Department of Biotechnology, Govt. of India [Grant No. BT/HRD/NBA/37/01/2014], and by Core Grant of National Institute of Plant Genome Research (NIPGR), New Delhi, India. Roshan K. Singh acknowledges the research fellowship received from Council of Scientific and Industrial Research, Govt. of India.

References

Austin D (2006) Fox-tail millets (Setaria: Poaceae)-abandoned food in two hemispheres. Econ Bot 60:143–158

Baltensperger DD (1996) Foxtail and proso millet. In: Janick J (ed) Progress in new crops. ASHS Press, Alexandria, VA, pp 182–190

Benabdelmouna A, Darmency H (2003) Copia-like retrotransposons in the genus *Setaria*: sequence heterogeneity, species distribution and chromosomal organization. Plant Syst Evol 237:127–136

Benabdelmouna A, Abirached-Darmency M, Darmency H (2001a) Phylogenetic and genomic relationships in *Setaria italica* and its close relatives based on the molecular diversity and chromosomal organization of 5S and 18S-5.8S-25S rDNA genes. Theor Appl Genet 103:668–677

Benabdelmouna A, Shi Y, Abirached-Darmency M, Darmency H (2001b) Genomic in situ hybridization (GISH) discriminates between the A and the B genomes in diploid and tetraploid Setaria species. Genome 44:685–690

Bennetzen JL, Schmutz J, Wang H, Percifield R, Hawkins J, Pontaroli AC, Estep M, Feng L, Vaughn JN, Grimwood J, Jenkins J, Barry K, Lindquist E, Hellsten U, Deshpande S, Wang X, Wu X, Mitros T, Triplett J, Yang X, Ye CY, Mauro-Herrera M, Wang L, Li P, Sharma M, Sharma R, Ronald PC, Panaud O, Kellogg EA, Brutnell TP, Doust AN, Tuskan GA, Rokhsar D, Devos KM (2012) Reference genome sequence of the model plant *Setaria*. Nat Biotechnol 30:555–561

Dekker J (2003) The foxtail (*Setaria*) species-group. Weed Sci 51:641–656

Devos KM, Wang ZM, Beales J, Sasaki T, Gale MD (1998) Comparative genetic maps of foxtail millet (*Setaria italica*) and rice (*Oryza sativa*). Theor Appl Genet 96:63–68

Diao X, Schnable J, Bennetzen J, Li J (2014) Initiation of *Setaria* as a model plant. Front Agr Sci Eng 1:16–20

Doebley JF, Gaut BS, Smith BD (2006) The molecular genetics of crop domestication. Cell 127:1309–1321

Doust AN, Devos KM, Gadberry MD, Gale MD, Kellogg EA (2004) Genetic control of branching in foxtail millet. Proc Natl Acad Sci USA 101:9045–9050

Doust AN, Devos KM, Gadberry MD, Gale MD, Kellogg EA (2005) The genetic basis for inflorescence variation between foxtail and green millet (poaceae). Genetics 169:1659–1672

Doust AN, Kellogg EA, Devos KM, Bennetzen JL (2009) Foxtail millet: a sequence-driven grass model system. Plant Physiol 149:137–141

Dwivedi S, Upadhyaya H, Senthilvel S, Hash C, Fukunaga K, Diao X, Santra D, Baltensperger D, Prasad M (2012) Millets: genetic and genomic resources. In: Janick J (ed) Plant breed reviews, vol 35. Wiley, USA, pp 247–375

Giussani LM, Cota-Sánchez JH, Zuloaga FO, Kellogg EA (2001) A molecular phylogeny of the grass subfamily Panicoideae (Poaceae) shows multiple origins of C4 photosynthesis. Am J Bot 88:1993–2012

Hunt HV, Vander Linden M, Liu X, Motuzaite-Matuzeviciute G, Colledge S, Jones MK (2008) Millets across Eurasia: chronology and context of early records of the genera *Panicum* and *Setaria* from archaeological sites in the old world. Veg Hist Archaeobot 17:5–18

Jusuf M, Pernes J (1985) Genetic variability of foxtail millet (*Setaria italica* (L.) P. Beauv): Electrophoretic study of five isoenzyme systems. Theor Appl Genet 71:385–391

Kole C, Muthamilarasan M, Henry R, Edwards D, Sharma R, Abberton M, Batley J, Bentley A, Blakeney M, Bryant J, Cai H, Cakir M, Cseke LJ, Cockram J, de Oliveira AC, De Pace C, Dempewolf H, Ellison S, Gepts P, Greenland A, Hall A, Hori K, Hughes S, Humphreys MW, Iorizzo M, Ismail AM, Marshall A, Mayes S, Nguyen HT, Ogbonnaya FC, Ortiz R, Paterson AH, Simon PW, Tohme J, Tuberosa R, Valliyodan B, Varshney RK, Wullschleger SD, Yano M, Prasad M (2015) Application of genomics-assisted breeding for generation of climate resilient crops: progress and prospects. Front Plant Sci 6:563

Lata C, Bhutty S, Bahadur RP, Majee M, Prasad M (2011) Association of an SNP in a novel DREB2-like gene SiDREB2 with stress tolerance in foxtail millet [*Setaria italica* (L.)]. J Exp Bot 62:3387–3401

Lata C, Gupta S, Prasad M (2013) Foxtail millet: a model crop for genetic and genomic studies in bioenergy grasses. Crit Rev Biotechnol 33:328–343

Lata C, Mishra AK, Muthamilarasan M, Bonthala VS, Khan Y, Prasad M (2014) Genome-wide investigation and expression profiling of AP2/ERF transcription factor superfamily in foxtail millet (*Setaria italica* L.). PLoS ONE 9:e113092

Le Thierry d'Ennequin M, Panaud O, Toupance B, Sarr A (2000) Assessment of genetic relationships between *Setaria italica* and its wild relative *S. viridis* using AFLP markers. Theor Appl Genet 100:1061–1066

Li P, Brutnell TP (2011) *Setaria viridis* and *Setaria italica*, model genetic systems for Panicoid grasses. J Exp Bot 62:3031–3037

Liang S, Yang G, Ma Y (2010) Chemical characteristics and fatty acid profile of foxtail millet bran oil. J Am Oil Chem Soc 87:63–67

Lu H, Zhang J, Liu KB, Wu N, Li Y, Zhou K, Ye M, Zhang T, Zhang H, Yang X, Shen L, Xu D, Li Q (2009) Earliest domestication of common millet (*Panicum miliaceum*) in East Asia extended to 10,000 years ago. Proc Natl Acad Sci USA 106:7367–7372

Mishra AK, Puranik S, Bahadur RP, Prasad M (2012) The DNA-binding activity of an AP2 protein is involved in transcriptional regulation of a stress-responsive gene, SiWD40, in foxtail millet. Genomics 100:252–263

Mishra AK, Muthamilarasan M, Khan Y, Parida SK, Prasad M (2014) Genome-wide investigation and expression analyses of WD40 protein family in the model plant foxtail millet (*Setaria italica* L.). PLoS ONE 9:e86852

Muthamilarasan M, Prasad M (2015) Advances in *Setaria* genomics for genetic improvement of cereals and bioenergy grasses. Theor Appl Genet 128:1–14

Muthamilarasan M, Bonthala VS, Mishra AK, Khandelwal R, Khan Y, Roy R, Prasad M (2014a) C_2H_2 type of zinc finger transcription factors in foxtail millet

define response to abiotic stresses. Funct Integr Genomics 14:531–543
Muthamilarasan M, Khandelwal R, Yadav CB, Bonthala VS, Khan Y, Prasad M (2014b) Identification and molecular characterization of MYB transcription factor superfamily in C_4 model plant foxtail millet (*Setaria italica* L.). PLoS ONE 9:e109920
Muthamilarasan M, Bonthala VS, Khandelwal R, Jaishankar J, Shweta S, Nawaz K, Prasad M (2015) Global analysis of WRKY transcription factor superfamily in Setaria identifies potential candidates involved in abiotic stress signaling. Front Plant Sci 6:910
Muthamilarasan M, Dhaka A, Yadav R, Prasad M (2016a) Exploration of millet models for developing nutrient rich graminaceous crops. Plant Sci 242:89–97
Muthamilarasan M, Mangu VR, Zandkarimi H, Prasad M, Baisakh N (2016b) Structure, organization and evolution of ADP-ribosylation factors in rice and foxtail millet, and their expression in rice. Sci Rep 6:24008
Oelke EA, Oplinger ES, Putnam DH, Durgan BR, Doll JD, Undersander DJ (1990) Millets. In: Alternative field crops manual, University of Wisconsin-Exension, Cooperative Extension
Puranik S, Bahadur RP, Srivastava PS, Prasad M (2011) Molecular cloning and characterization of a membrane associated NAC family gene, SiNAC from foxtail millet [*Setaria italica* (L.) P. Beauv.]. Mol Biotechnol 49:138–150
Puranik S, Sahu PP, Mandal SN, Venkata Suresh B, Parida SK, Prasad M (2013) Comprehensive genome-wide survey, genomic constitution and expression profiling of the NAC transcription factor family in foxtail millet (*Setaria italica* L.). PLoS One 8:e64594
Reddy VG, Upadhyaya H, Gowda C (2006) Characterization of world's foxtail millet germplasm collections for morphological traits. Int Sorghum Millets Newsl 47:107–109
Sema A, Sarita S (2002) Suitability of millet-based food products for diabetics. J Food Sci Technol (Mysore) 39:423–426
Singh RK, Jaishankar J, Muthamilarasan M, Shweta S, Dangi A, Prasad M (2016) Genome-wide analysis of heat shock proteins in C_4 model, foxtail millet identifies potential candidates for crop improvement under abiotic stress. Sci Rep 6:32641
Sivaraman L, Ranjekar PK (1984) Novel molecular features of millet genomes. Indian J Biochem Biophys 21:299–303
Thathola A, Srivastava S, Singh G (2010) Effect of foxtail millet (*Setaria italica*) supplementation on serum glucose, serum lipids and glycosylated hemoglobin in type 2 diabetics. Diabetologia Croat 40:23–28
Vavilov NI (1926) Studies on the origin of cultivated plants. Inst Appl Bot Plant Breed 16:1–248
Vicentini A, Barber JC, Aliscioni SS, Giussani LM, Kellogg EA (2008) The age of the grasses and clusters of origins of C4 photosynthesis. Glob Change Biol 14:2963–2977
Wang R, Wendel JF, Dekker JH (1995) Weedy adaptation in Setaria spp. I. Isozyme analysis of genetic diversity and population genetic structure in *Setaria viridis*. Am J Bot 82(3):308–317
Wang Y, Zhi H, Li W, Li H, Wang Y, Huang Z, Diao X (2009) A novel genome of C and the first autotetraploid species in the Setaria genus identified by genomic in situ hybridization. Genet Resour Crop Evol 56:843–850
Yadav CB, Muthamilarasan M, Pandey G, Prasad M (2015) Identification, characterization and expression profiling of Dicer-like, Argonaute and RNA-dependent RNA polymerase gene families in foxtail millet. Plant Mol Biol Rep 33:43–55
Yadav CB, Muthamilarasan M, Dangi A, Shweta S, Prasad M (2016) Comprehensive analysis of SET domain gene family in foxtail millet identifies the putative role of *SiSET14* in abiotic stress tolerance. Sci Rep 6:32621
Yang X, Wan Z, Perry L, Lu H, Wang Q, Zhao C, Li J, Xie F, Yu J, Cui T, Wang T, Li M, Ge Q (2012) Early millet use in northern China. Proc Natl Acad Sci USA 109:3726–3730
Zhang C, Zhang H, Li JX (2007) Advances of millet research on nutrition and application. J Chin Cereals Oils Assoc 22:51–55
Zhang G, Liu X, Quan Z, Cheng S, Xu X, Pan S, Xie M, Zeng P, Yue Z, Wang W, Tao Y, Bian C, Han C, Xia Q, Peng X, Cao R, Yang X, Zhan D, Hu J, Zhang Y, Li H, Li H, Li N, Wang J, Wang C, Wang R, Guo T, Cai Y, Liu C, Xiang H, Shi Q, Huang P, Chen Q, Li Y, Wang J, Zhao Z, Wang J (2012) Genome sequence of foxtail millet (*Setaria italica*) provides insights into grass evolution and biofuel potential. Nat Biotechnol 30:549–554
Zuloaga FO, Morrone O, Giussani LM (2000) A cladistic analysis of the Paniceae: a preliminary approach. In: Jacobs SWL, Everett J (eds) Grasses systematics and evolution CSIRO. Victoria, Australia, Collingwood, pp 123–135

Foxtail Millet Genome Sequencing, Assembly, Annotation, and Application

2

Mehanathan Muthamilarasan, Shweta Shweta and Manoj Prasad

Abstract

Among millets, foxtail millet (*Setaria italica* L.) is a well-studied crop, which is known for its genetic close-relatedness to biofuel grasses and cereals, C_4 photosynthesis, and appreciable tolerance to broad-spectrum abiotic stresses. Foxtail millet, along with its wild ancestor, green foxtail (*S. viridis* L.), are accentuated as model crops for studying the aforementioned traits. In view of their importance, the genomes have been sequenced and released. The present chapter summarizes the sequencing efforts, the outcomes, and the application of sequence information in expediting genomics studies. In addition, the chapter also provides a snapshot of how the genome sequence information has been exploited to develop different genomic resources useful for crop improvement.

2.1 Introduction

The term 'millets' denotes the small-seeded, polyphyletic, annual C_4 grasses belonging to the Poaceae family. This includes several species, including foxtail millet (*Setaria italica*), green foxtail (*Setaria viridis*), pearl millet (*Pennisetum glaucum*), finger millet (*Elusine coracana*), proso millet (*Panicum miliaceum*), and other minor millets which are farmed as food and fodder crops on deprived and trivial areas in drylands of temperate, subtropical, and tropical sections across the world (Dwivedi et al. 2011; Lata et al. 2013). Millets stand fifth in global grain production after rice, wheat, maize and sorghum (Dwivedi et al. 2011; Lata et al. 2013); however, they are not as popular as chief staple crops such as rice, maize, and wheat. Of note, millet species are a prime resource of energy for many civilizations, especially in developing countries, and it is still being consumed as a primary source of energy by considerable global population inhabiting the arid and semi-arid regions of the world (Muthamilarasan and Prasad 2015). Millets are referred to as species from not less than four distinct tribes of

M. Muthamilarasan (✉) · S. Shweta (✉) ·
M. Prasad (✉)
National Institute of Plant Genome Research (NIPGR), Aruna Asaf Ali Marg, New Delhi 110067, India
e-mail: manoj_prasad@nipgr.ac.in

M. Prasad (ed.), *The Foxtail Millet Genome*, Compendium of Plant Genomes,
DOI 10.1007/978-3-319-65617-5_2

PACMAD grasses, namely Paniceae, Paspaleae, Cynodonteae, and Eragrostideae. Pearl millet, proso millet, foxtail millet, and green foxtail millet belong to tribe Paniceae. Irrespective of their distinct domestication history, C_4 photosynthesis is a common attribute shared by all the millets. In C_4 photosynthesis, unique bundle sheaths and mesophyll cells (Kranz anatomy) are utilized to concentrate CO_2 in the neighborhood of ribulose bisphosphate carboxylase/oxygenase. This mechanism decreases photorespiration and enhances water utilization efficiency in millets, especially during drought and heat stress. C_4 plants also have superior nitrogen utilization efficiency, and therefore need a lower nitrogen supply to attain photosynthetic rates equivalent to C_3 plants (Huang et al. 2016). These attributes of C_4 photosynthesis accentuate millets as climate-resilient crops as they can flourish in soils of low nutritional content, typically acclimatize to diverse environmental stresses, require minimal resources, and have significant nutritional values (Muthamilarasan and Prasad 2015; Muthamilarasan et al. 2016a).

Genomics has offered several tools and techniques to exploit an interesting genome to study the genetic factors that contribute to a trait-of-interest and transfer the genes/alleles/QTLs (quantitative trait locuses) to target crops for their improvement. In this context, millet genomes could be a repertoire of novel genes and alleles, although the polyploid and complex nature of their genomes obstructs large-scale genomic studies (Goron and Raizada 2015; Saha et al. 2016). As a first step, sequencing a small and simple genome among millets could facilitate the implementation of comparative genomics tools to unravel the genomic complexities of other millets. In this regard, foxtail millet (*S. italica* L.) has been chosen as a tractable model because of its small genome size ($\sim$512 Mb), inbreeding nature, low repetitive DNA content, and short life cycle (Doust et al. 2009; Li and Brutnell 2011; Lata et al. 2013; Diao et al. 2014; Muthamilarasan and Prasad 2015). In this context, two international efforts have been made to sequence the genome of foxtail millet along with its wild relative, green foxtail (*S. viridis*), and the draft genome sequence is made available to the global research community through public databases (Zhang et al. 2012; Bennetzen et al. 2012). The present chapter summarizes the factors which have accentuated foxtail millet and green foxtail (collectively called Setaria) as models, their genome sequencing efforts, the outcome of the projects, annotation data, comparative genome mapping against grass genomes, and the application of the genome sequence information.

2.2 Foxtail Millet and Green Foxtail as Prospective Models

Foxtail millet and its wild ancestor green foxtail belong to the genus *Setaria* under the subfamily Panicoideae, which comprises nearly 125 different cultivated and wild species. Both the grasses share significant genome-level synteny with several bioenergy grasses, including switchgrass (*Panicum virgatum*), napiergrass (*Pennisetum purpureum*), and pearl millet (*Pennisetum glaucum*) (Doust et al. 2009; Li and Brutnell 2011). The Setaria genomes are also evolutionarily close to major cereals including sorghum (*Sorghum bicolor*), maize (*Zea mays*), and rice (*Oryza sativa*) (Doust et al. 2009). Although maize and sorghum were initially considered as the model for biofuel research, the large genome size of maize along with paleopolyploid evolutionary history, the bigger stature of plants, and the longer life cycle have limited the applicability of these crops. The small genome size with comparatively less repetitive DNA along with small plant size have given Setaria the advantage to serve as excellent surrogate genomes for studying the genetics and genomics of bioenergy grasses (Diao et al. 2014; Muthamilarasan and Prasad 2015).

In addition, Setaria utilizes a C_4 mode of photosynthesis, where phosphoenolpyruvate carboxylase (PEPC) supports the instantaneous uptake of carbon dioxide and transports it to RuBisCO for photosynthesis. Therefore, being a C4 plant, Setaria possesses an elevated photosynthesis rate as compared to other C_3 plants, even in the presence of high light intensity and high temperature (Way et al. 2014). Instant quenching and release of carbon dioxide by PEPC do not need a continuous opening of stomata for a longer duration, and thus a lesser amount of water is lost by transpiration, consequently contributing to the efficient use of water (Lata et al. 2013). In this context, Setaria could be an ideal model for studying C_4 photosynthesis. Recently, the abiotic stress tolerance of foxtail millet has gained importance, and several functional genomics studies aimed at characterizing several stress-responsive genes including *NAC* (Puranik et al. 2011, 2013), *WD40* (Mishra et al. 2012, 2014), *AP2/ERF* (Lata et al. 2011, 2014), *C_2H_2 zinc finger* (Muthamilarasan et al. 2014a), *MYB* (Muthamilarasan et al. 2014b), *DCL*, *AGO*, and *RDR* (Yadav et al. 2015a), *WRKY* (Muthamilarasan et al. 2015a), and *ADP-ribosylation factors* (Muthamilarasan et al. 2016b) suggested their putative involvement in stress-responsive molecular machinery. Thus, Setaria could be a prospective model system to study C_4 photosynthesis, abiotic stress tolerance, and biofuel traits.

2.3 Genome Sequencing Projects

The release of the whole genome sequence of foxtail millet and green foxtail is considered as a milestone in Setaria genomics. The genome was sequenced by two independent groups led by Beijing Genome Initiative (BGI), China and the United States Department of Energy Joint Genome Institute (USDOE-JGI), USA (Zhang et al. 2012; Bennetzen et al. 2012). The summary of these two sequencing efforts is given in Table 2.1.

2.3.1 Beijing Genome Initiative

2.3.1.1 Sequencing Overview

The foxtail millet strain 'Zhang gu' was sequenced using whole genome shotgun combined with next-generation sequencing technology. DNA libraries of various sizes were constructed and sequenced using Illumina second-generation sequencers, which generated a raw data of 63.5 Gb (127× sequencing depth). Post filtering of raw data, ~40 Gb (80×) clean reads were used for genome assembly by SOAPdenovo. The contig N50 and scaffold N50 were 25.4 kb and 1.0 Mb, respectively, which showed that 90% of contigs and scaffolds were present in 16,903 contigs and 439 longest scaffolds, respectively. The total length of all scaffolds was estimated to be 423 Mb, with 28 Mb (6.6%) gaps, and the genome size was determined as ~490 Mb (Zhang et al. 2012). Comparing the telomeric repeats identified in sorghum with foxtail millet showed the abundant presence of 155 bp repeat elements (Zhang et al.

Table 2.1 Summary of the outcomes of foxtail millet genome sequencing efforts

Information	Beijing Genome Initiative, China (Zhang et al. 2012)	United States Department of Energy Joint Genome Institute (Bennetzen et al. 2012)
Accessions sequenced	Foxtail millet cv. 'Zhang gu' and 'A10'	Foxtail millet cv. 'Yugu1' and green foxtail cv. 'A10'
Platforms used	Illumina second-generation sequencer	ABI3730xl capillary sequencer; 454 FLX platform; Illumina Genome Analyzer II platform
Sequence data generated	~40 Gb	~4 Gb for 'Yugu1' and ~3.5 Gb for 'A10'
Genome coverage	~86% (81% excluding gaps)	~80%
Genome size	~423 Mb	~400 Mb
Total number of genes	38801 (~82% expressed)	24,000–29,000 expressed genes

2012). The repeat elements were identified in the foxtail millet genome, which showed that 46% of the total element is composed of different transposable elements (TEs), namely DNA transposons (39.7 Mb–9.4%; CACTA, *h*AT, Tourist, Helitron, Stowaway and other tandem repeats) and retroelements (133.6 Mb–31.6%; LTR/*Copia*, LTR/*Gypsy*, LTR/other, LINEs and SINEs).

2.3.1.2 Genetic Map

A widely cultivated photo-thermosensitive male-sterile line 'A2' was also sequenced using Illumina GA II platform ($\sim 10\times$ depth) and the reads were mapped onto the reference genome assembly. This identified a total of 542,322 single nucleotide polymorphisms (SNPs), 33,587 small insertion and deletions (InDels), and 10,839 structural variants (SVs) between the two foxtail millet strains. From this, 118 SNPs and 641 SVs were used to genotype an F_2 population comprising 480 individuals developed by crossing 'Zhang gu' and 'A2'. Of these, 751 markers were used to construct the genetic map covering 1,865 centimorgan (cM) in total. These markers clustered into nine linkage groups and the comparison of genetic and physical distances revealed that 33% of the genome was located in low-recombination regions.

2.3.1.3 Foxtail Millet Genes

An integrated annotation pipeline was used to identify 38,801 genes in foxtail millet. The genes were mapped against several protein databases including SwissProt, TrEMBL InterPro, KEGG, and GO to retrieve the homologs with known function. This information was used to assign the function of foxtail millet genes, and a total of 30,579 genes (78.81%) were annotated whereas 8,220 (21.19%) were unannotated. The transcriptome of four tissues, namely root, leaf, spica, and stem, was sequenced and aligned to the genome assembly. The analysis showed that 81.7% of the predicted genes were expressed in these tissues. Average lengths of intron and exon of these genes were 442 and 256 bp, respectively, and the average exon per gene is 4.3 (Zhang et al. 2012). Gene ontology annotation of identified genes revealed that 78.8% of the genes have homologs with defined functions in the public database. In addition, 1,367 pseudogenes were predicted, which could be duplicated (both parent genes and pseudogenes are multi-exon genes), retrotransposed (parent genes are multi-exon genes whereas pseudogenes are single-exon genes), or unclassified (both parent genes and pseudogenes are single-exon genes or pseudogenes are multi-exon genes). Prediction of non-coding RNA genes showed the presence of 159 miRNA genes, 382 small nuclear RNA genes (CD-box, HACA-box, and splicing), 704 tRNA genes, and 99 rRNA genes (18S, 28S, 5.8S, and 5S) in foxtail millet genome. Large clusters of rRNA genes were found on chromosomes 1, 7, 8, and 9, whereas other non-coding RNA genes showed biased chromosomal distribution (Zhang et al. 2012).

2.3.1.4 Data Availability

The sequence reads of both genome and transcriptome are available in NCBI Sequence Read Archive (https://www.ncbi.nlm.nih.gov/sra?term=SRA048234) and DDBJ (https://trace.ddbj.nig.ac.jp/DRASearch/submission?acc=SRA048234). The genome sequence and annotation data set are available in NCBI (Project ID: PRJNA77795 (chromosomes) and PRJNA73995 (scaffolds); the accession number is GSM892310). The BGI team has also established a dedicated database for foxtail millet genome (http://foxtailmillet.genomics.org.cn), and the genome assembly, annotation, and other relevant data are available at ftp://ftp.genomics.org.cn/pub/Foxtail_millet.

2.3.2 United States Department of Energy Joint Genome Institute

2.3.2.1 Sequencing Overview

The USDOE-JGI has sequenced foxtail millet inbred line 'Yugu1' and green foxtail accession 'A10' using Sanger sequence analysis on ABI3730xl capillary sequencer and Illumina Genome Analyzer II platform, respectively. Also, 50688 BAC library clones ($\sim 12\times$ genome coverage) were also sequenced. The final genomic sequence of 'Yugu1' contained 396.7 Mb in nine chromosomes and 4.2 Mb in 327 scaffolds, which provided a genome coverage of ~80%. In the case of 'A10', ~3,500 Mb of generated data provided ~7× coverage (Bennetzen et al. 2012).

2.3.2.2 Genetic Map

A total of 247 progenies were generated by crossing foxtail millet inbred line 'B100' and green foxtail accession 'A10', and using this progeny, a recombinant inbred line (RIL) population was constructed (Bennetzen et al. 2012). This RIL population was sequenced using paired-end Illumina technology, and 3,149,093 SNPs were identified. Among these, 992 SNPs distributed at ~400-kb intervals were used to construct the genetic map comprising of nine linkage groups with a coverage of 1,416 cM. The genomes of foxtail millet and green foxtail were compared with sorghum and rice genomes, which revealed the presence of 188 and 163 syntenic blocks. These blocks were analyzed to identify the number of species-specific rearrangements in *Setaria* and sorghum relative to rice, which showed the presence of seven and four species-specific inversions in Setaria chromosomes 1, 2, 4, 6, 8, and 9 and the orthologous sorghum chromosomes relative to rice (Bennetzen et al. 2012).

2.3.2.3 Genome Annotation and Analysis

Annotation of fully assembled whole genome sequence predicted the presence of 35,472 protein-coding genes as primary transcripts and 5,128 are alternate transcripts. The analysis also showed that ~11% of these genes could be novel to foxtail millet and are interesting candidates to study. The average length of introns and exons of annotated genes were 163 and 135 bp, respectively, and the average length of proteins was 329 amino acids. These were in accordance with the findings reported in other grasses and Arabidopsis, which suggests a high degree of gene conservation among angiosperms (Bennetzen et al. 2012). In the case of transposable elements, 40% of 'Yugu1' genome is predicted to contain TEs which is relatively equal to rice (~40%) but lower than other grass genomes, including sorghum (~62%), maize, and wheat (more than 80%). The relatively lower content of TEs favors Setaria as a model. In addition, transcriptomes of different tissues and treatments were sequenced using different sequencing platforms, which showed 34,088 RNA-seq loci distributed on 9 chromosomes of foxtail millet, and 48 miRNA encoding genes were predicted. Similar to Zhang et al. (2012), the study also analyzed the *PEPC*, *PPDK*, and *MDH* genes among the sequenced grass genomes, which showed a high degree of conservation between maize and sorghum orthologs. The malic enzyme isoform of Setaria is different from those in maize and sorghum, which could be interesting to study for delineating the novel signatures of C_4 evolution. In addition, analysis of Setaria genomes identified six drought-associated gene clusters which possess more gene members in drought-tolerant species (Setaria and sorghum) as compared with drought-susceptible species (maize and rice) (Bennetzen et al. 2012; Muthamilarasan and Prasad 2017). These include plant lipid transfer protein, NADH oxidase, multi-antimicrobial extrusion protein, aldo/keto reductase, glutathione S-transferase, and AMP-dependent synthetase/ligase. The genes encoding for corresponding proteins could be potential candidates for further functional studies in foxtail millet and green foxtail to elucidate their genetic basis of stress adaptation.

2.3.2.4 Data Availability

The whole genome shotgun sequence reads of foxtail millet are available in the NCBI Nucleotide database under the accession number AGNK00000000 (https://www.ncbi.nlm.nih.gov/nuccore/AGNK00000000) and Phytozome (https://gold-dev.jgi.doe.gov/analysis_projects?id=Ga0039986). The updated version of the sequence reads are available under the accession number AGNK02000000, and consists of sequences AGNK02000001–AGNK02006778 (https://www.ncbi.nlm.nih.gov/Traces/wgs/?val=AGNK02). The scaffolds are available under the accession numbers CM003528–CM003536 and

KQ475381–KQ475707. EST and RNA-Seq data are available at NCBI EST database (accession numbers JK546897–JK608602) and Short Read Archive (SRX116346–SRX116357).

2.4 Comparative Mapping Between Foxtail Millet and Other Sequenced Grass Genomes

Foxtail millet genome was compared with the genome of other grasses, namely sorghum, maize, rice, and *Brachypodium*. A total of 24, 19, 20, and 29 large collinear blocks were identified between foxtail millet–rice, –sorghum, –*Brachypodium*, and –maize, respectively (Fig. 2.1). This indicated that foxtail millet genome is 71.8, 72.1, 61.5, and 86.7% collinear with rice, sorghum, *Brachypodium*, and maize genomes, respectively. This supports the close evolutionary relationships among the genomes, although the chromosomes have undergone extensive chromosomal shuffling (Zhang et al. 2012). The arrangement of conserved collinear blocks suggested the nested chromosomal fusions frequently observed in grass genomes. For example, foxtail millet chromosomes 2, 3, and 9 were collinear with rice chromosomes 7 and 9, 5 and 12, and 3 and 10, respectively, suggesting the divergence of these species from a common ancestor (Zhang et al. 2012). These three pairs of rice chromosomes were individually fused to form three chromosomes of foxtail millet and, among these three fusions, two also occurred in sorghum during evolution (Zhang et al. 2012). This suggests that the two fusions occurred before the divergence of foxtail millet and sorghum, around 27 million years ago.

In addition to genome-level synteny maps, several comparative maps have been constructed between the grass genomes at the level of markers and gene families (Muthamilarasan and Prasad 2015). The first comparative map using genomic SSRs (simple sequence repeats) was

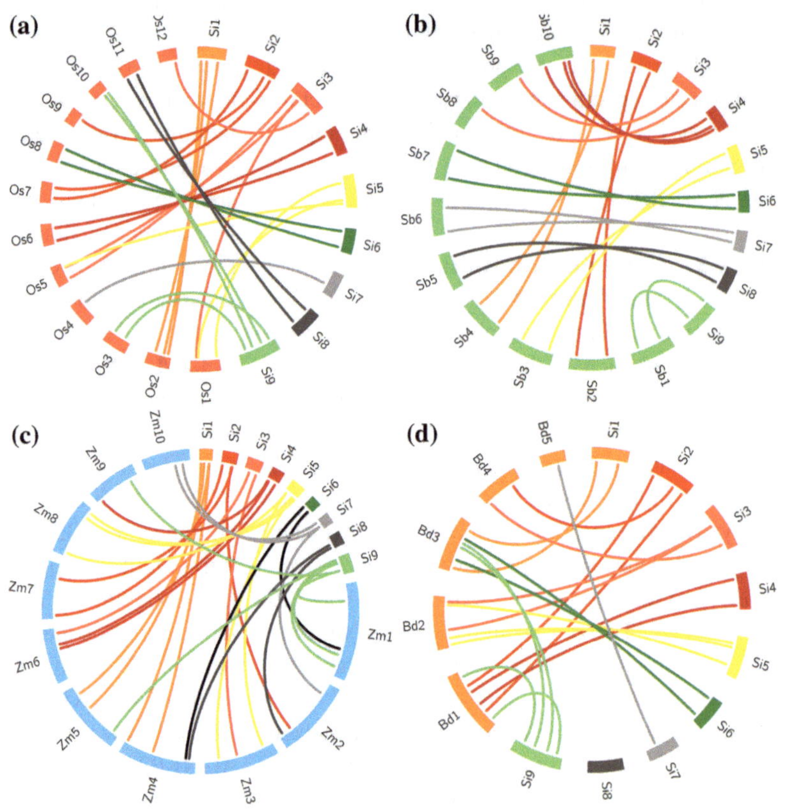

Fig. 2.1 Comparative mapping of conserved collinear blocks across the genomes of foxtail millet and **a** rice (Os; *Oryza sativa*), **b** sorghum (Sb; *Sorghum bicolor*), **c** maize (Zm; *Zea mays*), and **d** *Brachypodium* (Bd; *B. distachyon*)

performed by Pandey et al. (2013), and later by Kumari et al. (2013) using EST-derived SSRs. Later, Muthamilarasan et al. (2014a) developed comparative maps between foxtail millet and sequenced grass genomes using genome-wide intron-length polymorphic markers. Similarly, Yadav et al. (2014, 2015b) developed synteny maps using miRNA-based and transposable elements-based molecular markers. Comparative analysis of these maps revealed the percentage of sequence-based orthology and distribution of syntenic genomic regions between the genomes. Many of the physically mapped molecular markers in foxtail millet showed orthologous regions in more than one chromosome of other grass genomes. However, a higher percentage of synteny was observed between foxtail millet and sorghum, followed by foxtail millet and maize, in the case of all the molecular markers. Interestingly, it is unlike the comparative map constructed using collinear blocks, where maximum synteny was observed between foxtail millet and maize (86.7%; Zhang et al. 2012). These comparative maps also suggested several unique syntenic relationships between the genomes which were not reported by Devos (2005), Zhang et al. (2012), and Bennetzen et al. (2012) at genome level. The synteny maps also suggested the occurrence of nested chromosomal fusions frequently observed in grass genomes (Muthamilarasan and Prasad 2015). These marker-based comparative maps are very useful in the map-based isolation of genes of agronomic significance from foxtail millet using marker-based genotyping data of other related grass members.

Comparative maps were also constructed using the gene families responsible for stress response identified in foxtail millet. Members of gene families including *NAC* (Puranik et al. 2011, 2013), *WD40* (Mishra et al. 2012, 2014), *AP2/ERF* (Lata et al. 2011, 2014), C_2H_2 *zinc finger* (Muthamilarasan et al. 2014a), *MYB* (Muthamilarasan et al. 2014b), *DCL*, *AGO*, and *RDR* (Yadav et al. 2015a), *WRKY* (Muthamilarasan et al. 2015a), and *ADP-ribosylation factors* (Muthamilarasan et al. 2016b) were identified in foxtail millet to delineate their molecular and physiological roles in stress response, and comparative maps were constructed between foxtail millet and other grass genomes. Similar to marker-based maps, the gene-based maps also revealed the close evolutionary relationship between foxtail millet and sorghum, followed by maize and rice. The decrease in synteny between foxtail millet and rice as well as *Brachypodium* was expected because of the difference in the sub-families, where rice belongs to Ehrhartoideae and *Brachypodium* is a member of Pooideae. Muthamilarasan et al. (2015b) have identified and characterized the genes encoding proteins and enzymes which participate in cellulose (*CesA/Csl*), callose (*Gsl*), and monolignol biosynthesis, (*PAL*, *C4H*, *4CL*, *HCT*, *C3H*, *CCoAOMT*, *F5H*, *COMT*, *CCR*, and *CAD*) in foxtail millet. Comparative mapping of these genes on switchgrass genome revealed maximum homology between these two crops. This demonstrates the application of comparative maps in selecting candidate orthologous genes from related genomes for functional characterization.

2.5 Applications of Genome Sequence Information

The release of foxtail millet and green foxtail genomes have expedited the structural as well as functional genomics of both crops along with other millets, thus proving the efficient application of the sequence information in dissecting the agronomic traits and also facilitating crop improvement programs. This is evident from the increasing number of publications that arise on several aspects of Setaria (Fig. 2.2).

2.5.1 Molecular Marker Discovery

The release of whole genome sequences has facilitated the discovery of high-throughput molecular markers at genome-scale for large-scale genotyping applications. Zhang et al. (2012) have aligned the whole genome sequence of foxtail millet varieties 'Zhang gu' and 'A2' and identified 542,322 SNPs, 33,587 InDels, and 10,839 SVs. Bennetzen et al. (2012) have

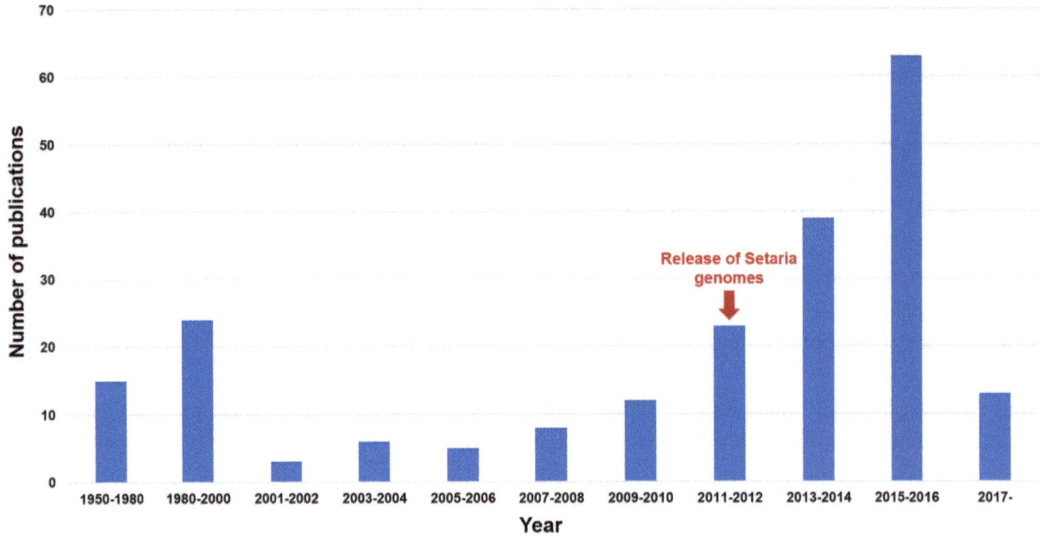

Fig. 2.2 Graphical representation of number of articles published on foxtail millet and green foxtail and indexed in NCBI PubMed

sequenced the RIL population from foxtail millet 'B100' X green foxtail 'A10' and identified 3,149,093 SNPs and used them for the construction of the genetic map. Pandey et al. (2013) have scanned the whole genome to identify 28,342 microsatellites and constructed a high-density physical map using 15,573 markers. Khan et al. (2014) have identified microRNAs from whole genome sequence information, which was then exploited to develop 176 miRNA-based markers (Yadav et al. 2014). Similarly, the genome sequence data was scanned to identify transposable elements, from which 20,278 repeat junction-based markers were developed (Yadav et al. 2015a, b). To build a haplotype map of foxtail millet, Jia et al. (2013) sequenced 916 foxtail millet varieties and identified more than 2.5 million SNPs. Through genome-wide association studies (GWAS), phenotyping of 916 varieties was carried out to identify 512 loci associated with 47 agronomic traits under five different environments. This large data set of loci with minor allele frequency >0.05 are linked with several essential agronomic traits in diverse environments. Similarly, Bai et al. (2013) performed resequencing of a foxtail millet landrace 'Shi-Li-Xiang' (SLX) and aligned the reads to both reference genomes to identify the patterns of genetic variations. The study demonstrated approximately 700,000 SNPs, 26,000 InDels 1–5 bp long, and 10,000 sequence variants between SLX and Yugu1 (Bennetzen et al. 2012); over 900,000 SNPs, 20,000 InDels, and 12,000 SVs between SLX and Zhang gu (Zhang et al. 2012; Bai et al. 2013). Recently, Upadhyaya et al. (2015) have resequenced 190 foxtail millet accessions including a core collection and identified 17,714 SNPs, which were then used for genetic structure analysis and GWAS. Altogether, the release of genome sequence data has facilitated the development of large-scale genome-wide molecular markers useful for high-throughput genotyping purposes.

2.5.2 Construction of High-Density Genetic and Physical Maps

Construction of genetic and physical maps followed by saturation approaches to pinpoint the genomic regions regulating agronomic traits is indispensable for molecular as well as genomics-assisted breeding. In the case of foxtail millet, genetic maps were constructed by Zhang et al. (2012), Bennetzen et al. (2012), Jia et al. (2013), Bai et al. (2013), and Upadhyaya et al. (2015), which were shown to be useful in mapping important traits. High-density physical maps were constructed by Pandey et al. (2013) using SSRs, Kumari et al. (2013) using EST-SSRs, Muthamilarasan et al. (Muthamilarasan et al. 2014c) using intron-length polymorphic markers, and Yadav et al. (2014, 2015a, b) using miRNA- and TE-based markers. In addition, gene family members including *NAC* (Puranik et al. 2011, 2013), *WD40* (Mishra et al. 2012, 2014), *AP2/ERF* (Lata et al. 2011, 2014), C_2H_2 *zinc finger* (Muthamilarasan et al. 2014a), *MYB* (Muthamilarasan et al. 2014b), *DCL*, *AGO*, and *RDR* (Yadav et al. 2015a), *WRKY* (Muthamilarasan et al. 2015a), and *ADP-ribosylation factors* (Muthamilarasan et al. 2016b) were also physically mapped onto foxtail millet chromosomes to construct physical maps. These maps are certainly useful in map-based selection of genes and/or markers for breeding and characteriszation purposes.

2.5.3 Gene Discovery

Whole genome sequence information provides access to the total number of genes that contribute to the growth, development and stress response. Increasing numbers of studies of individual genes and their corresponding gene families have suggested their involvement in diverse molecular, physiological and biological processes, and have provided novel clues on their regulation per se. In foxtail millet, *NAC* (Puranik et al. 2011, 2013), *WD40* (Mishra et al. 2012, 2014), *AP2/ERF* (Lata et al. 2011, 2014), C_2H_2 *zinc finger* (Muthamilarasan et al. 2014a), *MYB* (Muthamilarasan et al. 2014b), *DCL*, *AGO*, and *RDR* (Yadav et al. 2015a), *SET* (Yadav et al. 2016), and *Heat shock protein* (Singh et al. 2016) have been characterized at genome-wide level and interesting candidate genes showing differential expression in response to different abiotic stresses have been identified. In addition, a transcriptome-wide analysis of *WRKY* transcription factors was performed in green foxtail (Muthamilarasan et al. 2015a), and Muthamilarasan et al. (2016b) have also performed a comparative study on structure, organization, and evolution of *ADP-ribosylation factors* in rice and foxtail millet. Zhang et al. (2012) have untapped 586 genes in the genome which was annotated as 'response to water'. These genes could be candidates for studying the molecular machinery of drought and dehydration stress adaptation. Similarly, the foxtail millet-specific genes identified by Zhang et al. (2012) and Bennetzen et al. (2012) could also be studied for their specific roles in this climate-resilient crop.

2.5.4 Comparative Genomics

The availability of high-density genetic linkage maps as well as physical maps facilitates comparative mapping of genes and genomic regions of agronomic importance in other related crop species. The close evolutionary relationship between Setaria and several biofuel grasses as well as cereals and millets enables the transfer of genes, alleles, and QTLs from Setaria. It also facilitates the identification of orthologous genes in the related crop genome for characterizing their functional roles. The higher percentage of cross-genera transferability of molecular markers, namely SSRs, eSSRs, and ILPs, also paves the way for the use of these markers in other related genera whose genome sequence information is not available. In view of these, comparative physical maps were constructed in all the studies related to marker discovery and gene family identification in foxtail millet. In addition, comparative genomics also permits the evolutionary studies including rates of evolution of genes and their corresponding families, differential gene loss or retention following duplications and chromosomal rearrangements, and divergence patterns which collectively contribute to taxonomic, morphological, and physiological variations.

2.5.5 Sequencing and Resequencing of Genomes

The existence of predominant genome-level synteny between Setaria and other related species facilitates the sequencing and assembly of those complex genomes. Bennetzen et al. (2012) had sequenced the tetraploid switchgrass and demonstrated the applicability of foxtail millet genome in assembling the sequence reads. Availability of reference genomes also enables the resequencing of different cultivars, varieties, accessions, and landraces and alignment of sequence reads to the reference genome for identifying the SNPs and InDels which are useful for large-scale genotyping applications. In addition, reference genomes simplify the sequencing and analysis of transcriptomes. Unlike genomes, transcriptomes are highly dynamic, and assembly of RNA sequence reads using de novo approaches are often error prone. However, reference-based assembly allows an efficient assembly and annotation of transcripts, and thus the genome sequence plays an important role in these aspects.

2.6 Conclusions

The initiation of Setaria as model crops for dissecting the physiological, evolutionary and architectural traits of C_4 panicoid grasses was accentuated by the release of whole genome sequence information. This has accelerated the research in this neglected yet model crop, which promoted the crop from 'orphan' to 'crop with rich genetic and genomic resources'. It should be noted that foxtail millet research has now obtained numerous scientific leads to proceed further toward crop improvement. Attempts to develop genomic resources at large-scale and providing unrestricted access to the research community via web-based databases would certainly accelerate functional genomics studies and molecular breeding for crop improvement. Furthermore, the crop's potential abiotic stress tolerance has encouraged the plant research community to explore the respective molecular mechanism which would enable the generation of crops with improved stress tolerance, thus ensuring food security in the scenario of global climate change.

Acknowledgements Studies on millet genomics in Dr. Manoj Prasad's laboratory are supported by Science and Engineering Research Board (SERB), Department of Science and Technology (DST), Govt. of India [Grant No. EMR/2015/000464], by Department of Biotechnology, Govt. of India [Grant No. BT/HRD/NBA/37/01/2014], and by Core Grant of National Institute of Plant Genome Research (NIPGR), New Delhi, India. Shweta S. acknowledges the National Post-Doctoral Fellowship received from DST-SERB, Govt. of India [File No. PDF/2016/001430].

References

Bai H, Cao Y, Quan J, Dong L, Li Z, Zhu Y, Zhu L, Dong Z, Li D (2013) Identifying the genome-wide sequence variations and developing new molecular markers for genetics research by resequencing a Landrace cultivar of foxtail millet. PLoS ONE 8: e73514

Bennetzen JL, Schmutz J, Wang H, Percifield R, Hawkins J, Pontaroli AC, Estep M, Feng L, Vaughn JN, Grimwood J, Jenkins J, Barry K, Lindquist E, Hellsten U, Deshpande S, Wang X, Wu X, Mitros T, Triplett J, Yang X, Ye CY, Mauro-Herrera M, Wang L, Li P, Sharma M, Sharma R, Ronald PC, Panaud O, Kellogg EA, Brutnell TP, Doust AN, Tuskan GA, Rokhsar D, Devos KM (2012) Reference genome sequence of the model plant *Setaria*. Nat Biotechnol 30:555–561

Devos KM (2005) Updating the crop circle. Curr Opin Plant Biol 8:155–162

Diao X, Schnable J, Bennetzen JL, Li J (2014) Initiation of setaria as a model plant. Front Agri Sci Eng 1:16–20

Doust AN, Kellogg EA, Devos KM, Bennetzen JL (2009) Foxtail millet: a sequence-driven grass model system. Plant Physiol 149:137–141

Dwivedi S, Upadhyaya H, Senthilvel S, Hash C, Fukunaga K, Diao X, Santra D, Baltensperger D, Prasad M (2011) Millets: genetic and genomic resources. Plant Breed Rev 35:247–375

Goron TL, Raizada MN (2015) Genetic diversity and genomic resources available for the small millet crops to accelerate a New Green Revolution. Front Plant Sci 6:157

Huang P, Pu H, Studer AJ, Schnable JC, Kellogg EA, Brutnell TP (2016) Cross species selections can identify components of C4 photosynthesis in the grasses. J Exp Bot 68:127–135

Jia G, Huang X, Zhi H, Zhao Y, Zhao Q, Li W, Chai Y, Yang L, Liu K, Lu H, Zhu C, Lu Y, Zhou C, Fan D, Weng Q, Guo Y, Huang T, Zhang L, Lu T, Feng Q, Hao H, Liu H, Lu P, Zhang N, Li Y, Guo E, Wang S, Wang S, Liu J, Zhang W, Chen G, Zhang B, Li W, Wang Y, Li H, Zhao B, Li J, Diao X, Han B (2013) A haplotype map of genomic variations and genome-wide association studies of agronomic traits in foxtail millet (*Setaria italica*). Nat Genet 45:957–961

Khan Y, Yadav A, Suresh BV, Muthamilarasan M, Yadav CB, Prasad M (2014) Comprehensive genome-wide identification and expression profiling of foxtail millet [*Setaria italica* (L.)] miRNAs in response to abiotic stress and development of miRNA database. Plant Cell Tissue Organ Cult 118:279–292

Kumari K, Muthamilarasan M, Misra G, Gupta S, Subramanian A, Parida SK, Chattopadhyay D, Prasad M (2013) Development of eSSR-markers in *Setaria italica* and their applicability in studying genetic diversity, cross-transferability and comparative mapping in millet and non-millet species. PLoS ONE 8:e67742

Lata C, Bhutty S, Bahadur RP, Majee M, Prasad M (2011) Association of an SNP in a novel DREB2-like gene SiDREB2 with stress tolerance in foxtail millet [*Setaria italica* (L.)]. J Exp Bot 62:3387–3401

Lata C, Gupta S, Prasad M (2013) Foxtail millet: a model crop for genetic and genomic studies in bioenergy grasses. Crit Rev Biotechnol 33:328–343

Lata C, Mishra AK, Muthamilarasan M, Bonthala VS, Khan Y, Prasad M (2014) Genome-wide investigation and expression profiling of AP2/ERF transcription factor superfamily in foxtail millet (*Setaria italica* L.). PLoS ONE 9:e113092

Li P, Brutnell TP (2011) *Setaria viridis* and *Setaria italica*, model genetic systems for the Panicoid grasses. J Exp Bot 62:3031–3037

Mishra AK, Puranik S, Bahadur RP, Prasad M (2012) The DNA-binding activity of an AP2 protein is involved in transcriptional regulation of a stress-responsive gene, SiWD40, in foxtail millet. Genomics 100:252–263

Mishra AK, Muthamilarasan M, Khan Y, Parida SK, Prasad M (2014) Genome-wide investigation and expression analyses of WD40 protein family in the model plant foxtail millet (*Setaria italica* L.). PLoS ONE 9:e86852

Muthamilarasan M, Prasad M (2015) Advances in *Setaria* genomics for genetic improvement of cereals and bioenergy grasses. Theor Appl Genet 128:1–14

Muthamilarasan M, Prasad M (2017) Genetic determinants of drought stress tolerance in *Setaria*. In: Doust A, Diao X (eds) Genetics and genomics of Setaria. Springer, pp 267–289

Muthamilarasan M, Suresh BV, Pandey G, Kumari K, Parida SK, Prasad M (2014a) Development of 5123 intron-length polymorphic markers for large-scale genotyping applications in foxtail millet. DNA Res 21:41–52

Muthamilarasan M, Bonthala VS, Mishra AK, Khandelwal R, Khan Y, Roy R, Prasad M (2014b) C_2H_2-type of zinc finger transcription factors in foxtail millet define response to abiotic stresses. Funct Integr Genomics 14:531–554

Muthamilarasan M, Khandelwal R, Yadav CB, Bonthala VS, Khan Y, Prasad M (2014c) Identification and molecular characterization of MYB transcription factor superfamily in C_4 model plant foxtail millet (*Setaria italica* L.). PLoS ONE 9:e109920

Muthamilarasan M, Bonthala VS, Khandelwal R, Jaishakar J, Shweta S, Nawaz K, Prasad M (2015a) Global analysis of WRKY transcription factor superfamily in *Setaria* identifies potential candidates involved in abiotic stress signaling. Front Plant Sci 6:910

Muthamilarasan M, Khan Y, Jaishankar J, Shweta S, Lata C, Prasad M (2015b) Integrative analysis and expression profiling of secondary cell wall genes in C_4 biofuel model *Setaria italica* reveals targets for lignocellulose bioengineering. Front Plant Sci 6:965

Muthamilarasan M, Dhaka A, Yadav R, Prasad M (2016a) Exploration of millet models for developing nutrient rich graminaceous crops. Plant Sci 242:89–97

Muthamilarasan M, Mangu VR, Zandkarimi H, Prasad M, Baisakh N (2016b) Structure, organization and evolution of ADP-ribosylation factors in rice and foxtail millet, and their expression in rice. Sci Rep 6:24008

Pandey G, Misra G, Kumari K, Gupta S, Parida SK, Chattopadhyay D, Prasad M (2013) Genome-wide development and use of microsatellite markers for large-scale genotyping applications in foxtail millet [*Setaria italica* (L.)]. DNA Res 20:197–207

Puranik S, Bahadur RP, Srivastava PS, Prasad M (2011) Molecular cloning and characterization of a membrane associated NAC family gene, SiNAC from foxtail millet [*Setaria italica* (L.) P. Beauv.]. Mol Biotechnol 49:138–150

Puranik S, Sahu PP, Mandal SN, Venkata Suresh B, Parida SK, Prasad M (2013) Comprehensive genome-wide survey, genomic constitution and expression profiling of the NAC transcription factor family in foxtail millet (*Setaria italica* L.). PLoS One 8:e64594

Saha D, Dipnarayan S, Gowda MVC, Lalit A, ManjushaV Bansal KC (2016) Genetic and genomic resources of small millets. Crit Rev Plant Sci 35:56–79

Singh RK, Jaishankar J, Muthamilarasan M, Shweta S, Dangi A, Prasad M (2016) Genome-wide analysis of heat shock proteins in C_4 model, foxtail millet identifies potential candidates for crop improvement under abiotic stress. Sci Rep 6:32641

Upadhyaya HD, Vetriventhan M, Deshpande SP, Sivasubramani S, Wallace JG, Buckler ES, Hash CT, Ramu P (2015) Population genetics and structure of a global foxtail millet germplasm collection. Plant Genome 8:3

Way DA, Katul GG, Manzoni S, Vico G (2014) Increasing water use efficiency along the C_3 to C_4 evolutionary pathway: a stomatal optimization perspective. J Exp Bot 65:3683–3693

Yadav CB, Muthamilarasan M, Pandey G, Prasad M (2014) Development of novel microRNA-based genetic markers in foxtail millet for genotyping applications in related grass species. Mol Breed 34:2219–2224

Yadav CB, Muthamilarasan M, Pandey G, Prasad M (2015a) Identification, characterization and expression profiling of Dicer-like, Argonaute and RNA-dependent RNA polymerase gene families in foxtail millet. Plant Mol Biol Rep 33:43–55

Yadav CB, Bonthala VS, Muthamilarasan M, Pandey G, Khan Y, Prasad M (2015b) Genome-wide development of transposable elements-based markers in foxtail millet and construction of an integrated database. DNA Res 22:79–90

Yadav CB, Muthamilarasan M, Dangi A, Shweta S, Prasad M (2016) Comprehensive analysis of SET domain gene family in foxtail millet identifies the putative role of *SiSET14* in abiotic stress tolerance. Sci Rep 6:32621

Zhang G, Liu X, Quan Z, Cheng S, Xu X, Pan S, Xie M, Zeng P, Yue Z, Wang W, Tao Y, Bian C, Han C, Xia Q, Peng X, Cao R, Yang X, Zhan D, Hu J, Zhang Y, Li H, Li H, Li N, Wang J, Wang C, Wang R, Guo T, Cai Y, Liu C, Xiang H, Shi Q, Huang P, Chen Q, Li Y, Wang J, Zhao Z, Wang J (2012) Genome sequence of foxtail millet (*Setaria italica*) provides insights into grass evolution and biofuel potential. Nat Biotechnol 30:549–554

Transposable Elements in *Setaria* Genomes

3

Chandra Bhan Yadav and Manoj Prasad

Abstract

Large fractions of millet genomes are saturated with repetitive elements in which the major portion of repeats are contributed by transposable elements (TEs) which are also known as selfish genetic elements. These elements are capable of mobilizing in the genome from one position to another through transposition or retro-transposition with the help of certain specific enzymes coded by these TEs themselves. Recent advances in genome sequencing and assembly techniques provide an opportunity to enlighten our views on the current understanding of millet TE diversity and evolution in the genome. Transposable elements (TEs) represent approximately 40% of assembled millet genomes, and deeply branching lineages such as rice, maize, and other grass genomes exhibit a higher TE diversity in comparison to other plant taxa. With the advancement of sequencing techniques and availability of assembled genomes, long-read sequencing should soon provide access to TE-rich genomic regions of TE and their architecture in the genome. Furthermore, the current bottleneck in genome analyses and annotation of TEs could also be resolved to avoid misleading conclusions on repeat architecture and their involvement in genome evolution.

3.1 Introduction

Millets are the oldest cereals in the world and are thought to have been domesticated in China, Asia, and other parts of the world, including India. Millets have many excellent characteristics such as, being C_4, it possesses higher photosynthetic rates compared to other C_3 plants (Muthamilarasan and Prasad 2015). Most importantly, they have more water use efficiency

C.B. Yadav · M. Prasad (✉)
National Institute of Plant Genome Research (NIPGR), Aruna Asaf Ali Marg, New Delhi 110067, India
e-mail: manoj_prasad@nipgr.ac.in

M. Prasad (ed.), *The Foxtail Millet Genome*, Compendium of Plant Genomes,
DOI 10.1007/978-3-319-65617-5_3

in comparison to other related members of grass families such as maize, wheat, and sorghum. Millet grains are also most valuable for their high nutritional contents such as protein, folic acid, vitamins, and carotenoids (Muthamilarasan et al. 2016). Millet is a diverse group of annual crops from the family Poaceae that distinctively produces small seeds. It basically includes several items of food and pasturage, namely foxtail millet (*Setaria italica*), pearl millet (*Pennisetum glaucum*), finger millet (*Eleusine coracana*), proso millet (*Panicum miliaceum*), little millet (*Panicum sumatrense*), barnyard millet [*Echinochloa crus-galli* (Japanese) and *Echinochloa colona* (Indian)], kodo millet (*Paspalum scrobiculatum*), tef (*Eragrostis tef*), and guinea millet (*Brachiaria deflexa*) (Dwivedi et al. 2012). During the period 2004–2008 the average contribution of total cereal to world food production was 255.1 million tonnes per year, of which the millet share was 12.7% (32.3 million tonnes) (Dwivedi et al. 2012). These grasses are primarily adapted to the arid and semi-arid tropics' of Asia and Africa (particularly India, China, Mali, Nigeria, and Niger). India's contribution to world millet production is 32%. Furthermore, of the total millet produced, 90% is utilized in developing countries (FAO 1995). In terms of world millet production, foxtail millet and proso millet (*P. miliaceum*) rank second after pearl millet (*P. glaucum*) (FAOSTAT 2005; http://faostat.fao/org/).

In recent years, the availability of foxtail millet (*S. italica*) and green millet (*Setaria viridis*) genome sequences to the public provides an important resource for investigating plant genome architecture, drought tolerance, N_2 use efficiency, C_4 photosynthesis proficiency, and bioenergy crops (Bennetzen et al. 2012; Long 1999; Sage and Pearcy 2000; Kocacinar et al. 2008; Ghannoum et al. 2011). Both millets are now being used as model plants for experiments because of their small genome size, self-fertilization, and short life cycle. If stored properly, millets can keep well for 2 years or beyond without losing viability. Most millets, including foxtail millet, are advantageous for consumption as they are rich in nutrition, possess non-glutinous and non-acid forming properties, and are easy to digest. Moreover, the nutritional value of millets is high as they are rich in nutritionally important minerals such as iron, calcium, zinc, magnesium, phosphorus, and potassium. They are also a good source of dietary fiber and several vitamins (β-carotene, niacin, vitamin B6, and folic acid). The high lecithin content of millets is helpful for increasing strength of the nervous system. Therefore, consumption of millets can provide a way to eliminate malnutrition among a major section of the Indian population (Muthamilarasan et al. 2016). They are usually called coarse grains; however, because of their important nutritional properties they are now being termed 'nutria-millets/nutria-cereals'. Millets are also rich in polyphenols, tannins, and phytosterols, and are a good source of antioxidants. Despite possessing plentiful qualities, the use of millets as food is restricted to conventional consumers, predominantly the tribal populations. This can be attributed to the non-availability of user-friendly products such as ready-to-use/ready-to-eat millet-based food. In recent times, millets have been considered to be important food, and various efforts are under way to prepare suitable value-added ready-to-use products from them.

Based on reference genomes, a large number of studies have been conducted to investigate the genomic organization using DNA markers (SSRs (simple sequence repeats), EST-SSRs (expressed sequence tag-SSRs), intron-length polymorphic markers, SNPs (single nucleotide polymorphisms), InDels) (Pandey et al. 2013; Gupta et al. 2011, 2012; Muthamilarasan et al. 2014) and small RNA (ribonucleic acid)-based markers (Yadav et al. 2015). Many of the regulatory quantitative trait locuses (QTLs) and genes have been identified for various agronomic traits (Jia et al. 2013; Qie et al. 2014). Besides genes and other regulatory elements, there are other important substructures existing in the genome that could increase the complexity of the genome, and hence, considering their prominent role in biological regulations, repetitive elements are being explored to discover the genomic organization and the effect on phenotypic variations.

These repetitive sequences are randomly situated in the genome and have the ability to mobilize in the genome from one location to another. Later on, these mobile genetic elements were named transposable elements (TE), first discovered by Barbara McClintock who observed the existence of 'jumping genes or mobile genetic elements' in the maize genome, which are repetitive in nature. TEs prediction and identification of putative roles in regulating the various agronomic traits have been under way on the pilot plan in many eukaryotes as well as in almost all plant species genomes. TEs are small segments of the chromosome that can 'jump' out of their location into another completely random location (transposition). Because of their mobile behavior, the TEs are independently integrated throughout the genome, forming a major component of plant genomes and contributing to the diversity in genome sizes and structure. Here, we review recent insights into TEs in the era of high-throughput sequencing and the high-quality reference genome assembly and how it has influenced the millets' genomes complexity.

TEs are distinct DNA fragments that have the capability to organize themselves by mobilizing within the genome, either by transcribing new copies or replicating themselves during the process of translocation. The process of TE translocation within the genome would lead to reorganization of the genome and dramatically change the genome complexity over time. Earlier, TEs were considered to be functionless genomic parasites which are known as selfish genes. However, with the advancement of high-throughput sequencing and growing accessibility of assembled genomes has provided the opportunity to dissect out the complex role of TEs and their role in genome evolution (Yadav et al. 2015). Various researchers have also reported that TEs are important genetic elements which could have a crucial impact on genome architecture and evolution in numerous ways. Another interesting feature of regulation of genomic changes by TE is that they act as either alternate promoters or first exons for a subset of host genes, thereby bringing about changes in the rate of transcription/silencing (Peaston et al. 2004). We are therefore hypothesizing that various types of TEs could trigger sequential reprogramming of the genome during the different developmental stages. For example, it is reported that different LTR (long terminal repeat) retrotransposons have specific, developmentally regulated expression patterns, suggesting that the normal repressive chromatin structure for these loci might be established sequentially during megaspore mother cell differentiation to embryo development stages (Durán-Figueroa and Vielle-Calzada 2010).

Transposition of TEs from one position on the chromosome to another leads to phenotypic variation because of mutations caused by chromosomal breakage and fusion, as well as the transposition of genes. For example, in maize, Barbara McClintock identified two 'movable' elements—Activator (Ac) and Dissociation (Ds)—with the capability to insert into different genomic locations, either in genes or regulatory regions, leading to mutations in those genes and formation of new phenotypes. Therefore, researchers are using publicly available EST and genomic databases to identify the TEs in plant genomes and their involvement in regulating the various agronomic trait, either by inserting themselves into a gene or in other regulatory regions through mRNA (messenger RNA). Alternatively, searching for TE would be feasible in an RNA population using advanced RNA sequencing technologies. Most of the plant genome appears to be rich in the repetitive region that contains a high copy of retrotransposons, and that could be a most transcriptionally active region in the genome (Arabidopsis Genome Initiative 2000; The Rice Chromosome 10 Sequencing Consortium 2003). In certain LTR types of retrotransposons, RNA transcript is generated from Ty 1-copia-type and Ty3-gypsy-type retro elements. As these transcripts possess a poly tail A, they have the potential to be included in the mRNA populations. A search for TEs in the ESTs databases has been attempted by Macas et al. (2003) in *Pisum sativum*. They showed that at least a few elements are transcribed, most likely because of their association with genic regions. In another study, Echenique et al. (2002) used the

Triticeae ESTs database to assess the presence of TEs in the RNA population, such as retrotransposons (Ty 1-copia-type and Ty3-gypsy-type retro elements). The indirect estimation of transcriptional activity of these repetitive elements is important for improving the annotation of genomic sequences which are used to search these EST databases. Vicient (2010) analyzed the transcriptional activity of the maize TEs based on EST databases and found that 1.5% maize ESTs show sequence similarity with TEs. The large number of ESTs in the database could help in the identification of retrotransposons, which is the essential step for further studying expression analysis and miRNA-mediated silencing.

3.2 Classification of TEs

TEs have their own peculiar mechanism for translocations in the genome, either through RNA intermediates or without RNA intermediates. Based on the type of transposition and replication mechanisms, TEs are categorized into two classes: class I TEs [(retrotransposons produce RNA intermediates that are copied into DNA and then inserted into new locations within the genome by the action of reverse transcriptase (RT)] and class II TEs [DNA transposons move directly by a "cut and paste" mechanism] (Sabot et al. 2004). Class I elements are further classified into two superfamilies: (1) LTR-type retrotransposons, flanked by LTRs and (2) non-LTR elements [such as Long Interspersed Nuclear Elements (LINEs) and Short Interspersed Nuclear Elements (SINEs)].

Retrotransposons belong to this class, and use RNA intermediates to transpose and their replicative mode of transposition generates a new copy of transposon although the original transposon remains intact (Feschotte et al. 2002). Based on both their degree of sequence similarity and the order of encoded gene products, the class I LTR-type elements are further classified into Ty1/copia-like and Ty3/gypsy-like) (Fig. 3.1). During LTR retrotransposon mobilization, an

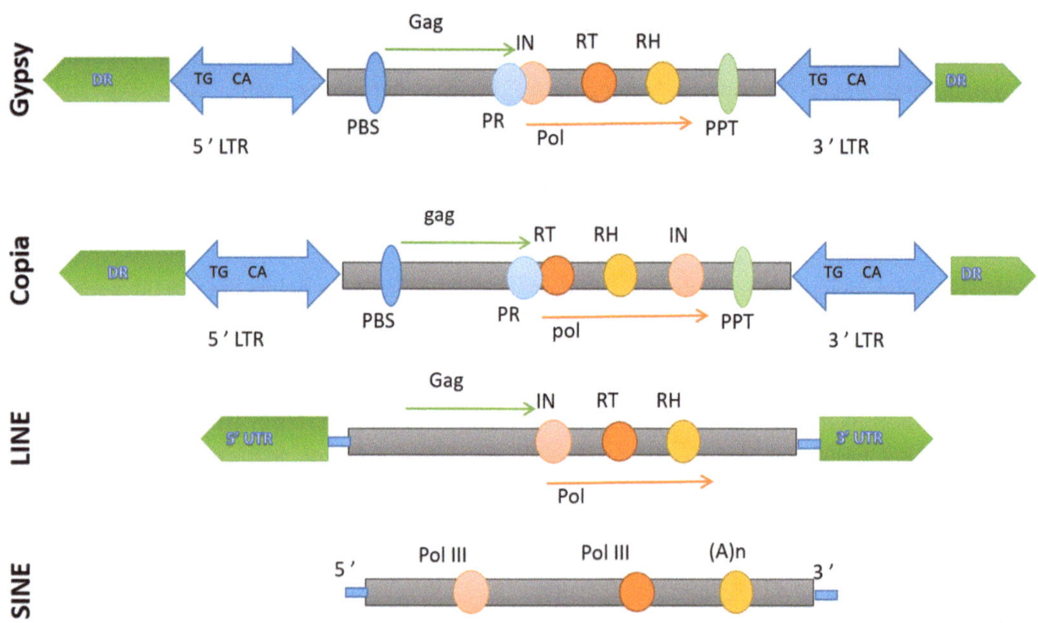

Fig. 3.1 Diagrammatic representation of major TE types and their classification scheme. Structural features are presented as LTR types (Copia and Gypsy) nucleocapsid protein (GAG), polyprotein (Pol), integrase (IN), reverse transcriptase (RT), and an RNase H (RH). LTR has 5′ and 3′ LTR flanking the internal coding region. Internal region of the LTR contains PBS and DR: represented as direct repeat. UTR is untranslated Region, Pol III: RNA Polymerase III promoter recognition sites, LINEs: Long Interspersed Nuclear Elements, SINEs: Short Interspersed Nuclear Elements

RNA molecule is reverse transcribed into double-stranded DNA through a series of template switches encoded by four genes containing protease, integrase, reverse transcriptase, and ribonuclease H. domains (Kumar and Bennetzen 1999). The double-stranded DNA copy of the element is then reinserted into the genome via an integrase. Full-length LTR flank two LTR named 5′ LTR and 3′ LTR started with TG and ended with CA flanking the internal coding region. The internal region of the LTR contains PBS (primer binding site) and polypurine tract (PPT) which were 20 and 15 bp in length, respectively. On the other hand, LINEs mobilize via a target-primed reverse transcription mechanism in which the transcript itself re-enters the nucleus with the help of its protein products. LINE contains an untranslated region (UTR) at both ends along with an RNA polymerase II promoter and two non-overlapping open reading frames (ORF1 and ORF2) encoding integrase, reverse transcriptase, and ribonuclease H. domains (Fig. 3.1). These proteins, usually including an endonuclease domain along with the reverse transcriptase, nick the target site for integration and conduct reverse transcription of the LINE transcript (Kumar and Bennetzen 1999). SINEs are also transposed through RNA intermediates by reverse-transcribed RNA molecules using RNA polymerase III (Fig. 3.1). The mechanism of retrotransposition of SINEs is slightly different and more complicated than LINEs. SINEs do not encode all the functional proteins involved in retrotransposition and therefore depend on other mobile elements for transposition. In some cases they may have their own endonuclease that allows them to cleave their way into the genome, but the majority of SINEs integrate at chromosomal breaks by using random DNA breaks to prime reverse transcriptase.

Class II TEs or DNA transposons mobilize without reverse transcription of source elements and are classified into three major subclasses, each with a distinct transposition mechanism: "cut-and-paste" or Terminal Inverted Repeat (TIR) DNA transposons (e.g., hATs, piggyBacs and mariners), rolling-circle transposons (e.g., Helitrons), and self-synthesizing DNA transposons. The transposase enzyme has a number of functional domains, including a catalytic domain and a DNA-binding domain, which catalyze the whole set of reactions necessary for DNA transposition (Fig. 3.2). Furthermore, they are classified into ten sub-classes, namely DNA/CMC-EnSpm, DNA/En-Spm, DNA/hAT-Ac, DNA/hAT-Tag1, DNA/hAT-Tip100, DNA/MULE-MuDR, DNA/PIF-Harbinger, DNA/TcMar-Stowaway, RC/Helitron, and DNA/Tourist based on similarity searches with known TEs reported in other plant species. TE contributes approximately half of the genomic region, and therefore it is important to identify the new class of retrotransposons that provide us with an unprecedented rate of discovery. This type of discovery provides evidence of the lineage-specific expansion of the specific class of TE that could certainly have influenced the evolution of the genome. The replicative mode of transposition by these retrotransposons results in their greater abundance in the genome, and reports indicate that the genomes of maize, *Arabidopsis*, and wheat are comprised of 50, 14, and 90% of LTR-type retrotransposons, respectively (SanMiguel et al. 1998; Arabidopsis Genome Initiative 2000; Meyers et al. 2001; Brenchley et al. 2012).

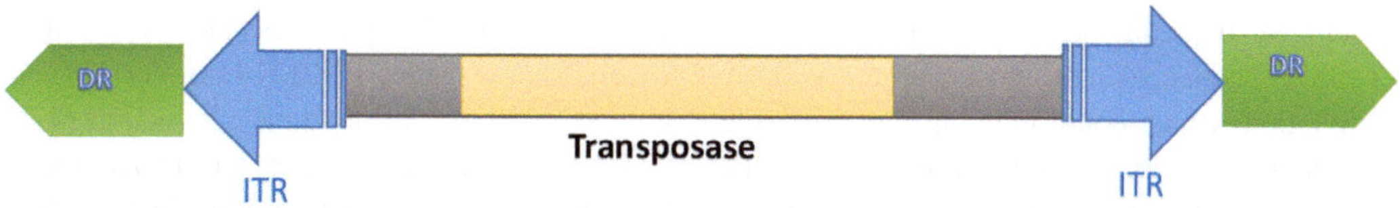

Fig. 3.2 DNA transposons having two terminal inverted repeats (ITRs) flanked by short direct repeats (DRs). It has a single ORF encoding transposase-like protein

3.3 Diversity and Evolution of TEs

TEs are usually distributed throughout the genome in the plant kingdom. Class I-type retrotransposons have been found to be widely distributed in the grass families, including major millets and non-millets. Furthermore, class I-type retrotransposons, especially Copia and Gypsy types of retrotransposons, are frequently found in high copy numbers in grass genome which makes the large and complex genome. Moreover, non-LTR retrotransposons, LINEs and SINEs are also an important component of the genome and randomly distributed in the genome. However, SINEs are present in low copy numbers in the millet genome as compared to other grasses as well as plant species. In general, retrotransposons are predominantly composed of noncoding sequences, present in an intron or an intergenic region of the plant genomes (Feschotte et al. 2002), as the replicative mode of retroelement transposition allows the LTR retrotransposon to accumulate in high copy numbers. Many of the studies revealed that plant genomes have substantial uniformity in the organization of TEs. For example, only ~18% of the Arabidopsis genome (The Arabidopsis Genome Initiative 2000) contributed to similarity with TE sequences, whereas up to 90% of the *Zea mays* (maize) and wheat genome is composed of TEs (Bennetzen and Kellogg 1997; Feschotte et al. 2002; Meyers et al. 2001; SanMiguel et al. 1998; Flavell et al. 1998). The transposition mechanism of TEs resulted in enriched heterochromatin formation through the process of chromatin modifications and DNA methylation, leading to transcriptional silencing. SanMiguel et al. (1996), Rabinowicz et al. (1999) reported that most of the TEs are limited to the methylated heterochromatin region in the maize genome. Vicient (2010) concluded that TEs and other repetitive sequences are most prone to DNA methylation, which could play an important role in structural and functional variation in plants. TEs/retrotransposons also play an important role in miRNA-mediated silencing. For example, a recent study demonstrated that small RNAs (sRNAs) of 24 nucleotides derived from the LTR region of TEs (TEs) interacted with AGO9 and induced TE silencing during the male and female gametophytes (Olmedo-Monfil et al. 2010).

3.4 TE Insertion in Millet Genes

Because of their mobile or jumping nature, TEs are supposed to be involved in disrupting the genes or integrating into nearby genes affecting the process of transcription. The phenomenon of translocation and integration of TEs into the genome leads to chromosomal rearrangements which may cause the accumulation of mutations through deletion or insertion mutation, and, ultimately, it affects the transposition process at a particular position toward the inactivation of genes. Yadav et al. (2015) detected 12% foxtail millet genes were found to be interrupted by TEs, of which ~0.75% genes had Copia-type retrotransposons, ~1% genes had Gypsy-type, ~3% genes had LINEs, and ~6% genes were interrupted with DNA transposons (Fig. 3.3). A similar phenomenon of TE insertion in the intronic region was also observed in green millet genome. However, 10% green millet genes were interrupted with TEs, in which ~0.54% genes had Copia-type retrotransposons, ~1% genes had Gypsy-type, ~1% genes had LINEs, and ~5% genes were interrupted with DNA transposons (Fig. 3.3).

TE integration in the genic region leads to suppression of gene expression. Functional annotation analysis of these TE interrupted genes revealed that major genes are involved in organ, reproductive development, and stress-responsive pathways. Further chromatin regulatory genes containing SET domain gene family and AGO gene family, which are the components of RNAi machinery, were identified as the genes interrupted with TEs. In a previous study, some of the foxtail millets genes (Si013141m, Si033830m, and Si009202m) have been characterized by Yadav et al. (2015) and they are interrupted by TE elements. Besides interruption in gene fragments, transposition events are used to provide a platform to capture gene fragment by many transposons and retroelements and become part

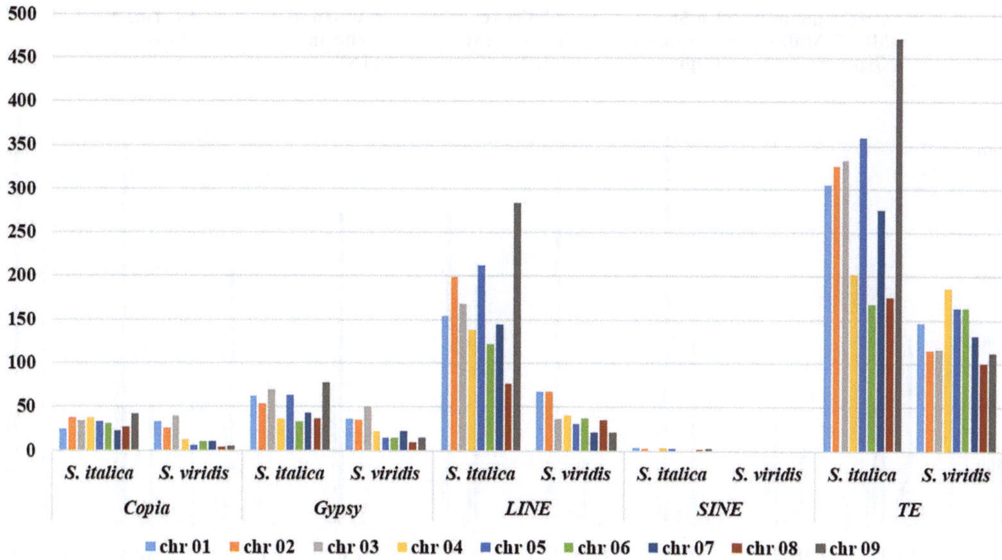

Fig. 3.3 Insertional pattern of different classes of TEs in the genic regions of foxtail millet (*Setaria italica*) and green millet genes (*Setaria viridis*)

of the gene regulatory regions at the site of insertion. Transposons pick up bits and pieces of genes that code for proteins other than transposases, and transposase genes are pressed into services other than transposition (Lisch and Bennetzen 2011; Lisch 2009). For example, the proteins encoded by the FAR1 and FHY3 genes of *Arabidopsis* are related to the MuDR family of transposases (Allen et al. 2006). FHY3 and FAR3 are transcription factors that regulate light-dependent chlorophyll biosynthesis in development, the former also gating phytochrome signaling to the circadian clock (Tang et al. 2012).

3.5 TE Patterns in the Millet Genome

Millets are herein well-defined as the paraphyletic group comprising major millets (*S. italica*, *S. viridis*, and *P. glaucum*), minor millets (Bandyopadhyay et al. 2017), and other related biofuel crops covering the deepest lineage of panicoids. In monocots, there are intense disparities in TE distribution and composition in different plant taxa, ranging from 90% in maize to 40% in rice. All major types of TEs are present in grass genome, and millets genome demonstrated overall higher TE diversity as compared to other members of grass family genomes.

3.5.1 *Setaria italica*

Information for various classes of TEs has been curated by Yadav et al. (2015) in foxtail millet, which is derived from whole genome sequencing. Publicly available foxtail millet genome assembly suggests that at least 40% of the genome is contributed by TEs. The distribution of TEs is consistent with other grass genomes such as rice (40%), and sorghum (62%) (Bennetzen et al. 2012). In silico curation by Yadav et al. (2015) suggested that most of the TEs are contributed by LTR-type retrotransposons followed by DNA transposons (class II). For example, Yadav et al. (2015) scanned the whole genome sequence of foxtail millet and identified a total of 1038 full-length Copia-type and 1570 full length Gypsy-type retrotransposons (Fig. 3.4). According to Yadav et al. (2015), the length of Copia elements and Gypsy-type LTRs varies greatly with a mean of 7.7 and 11.7 kb, respectively.

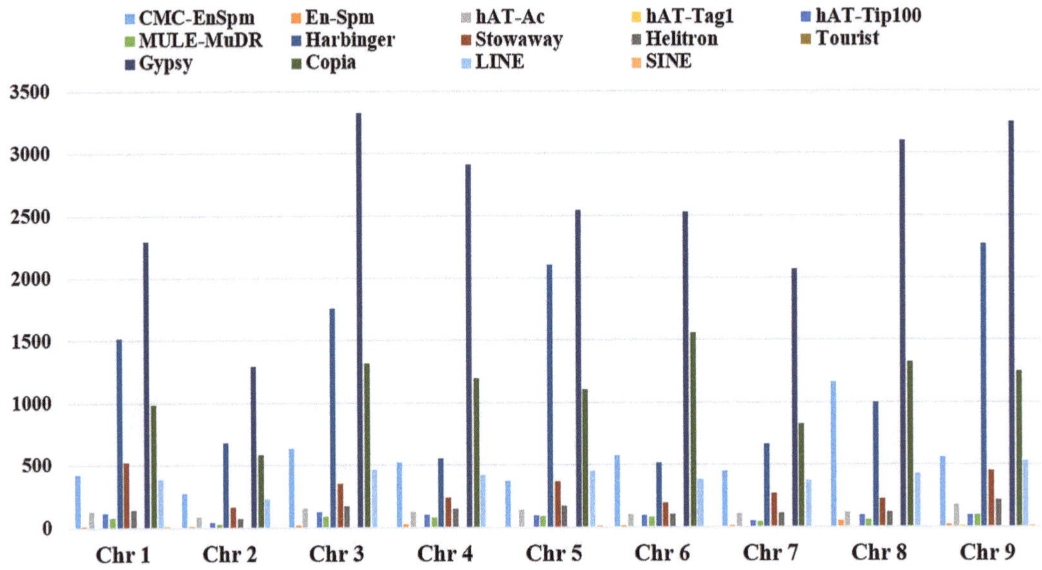

Fig. 3.4 Distribution of different classes of TEs across the nine chromosomes of foxtail millet

They have also curated a large number of non-LTRs elements which have been categorized as LINEs and SINEs. The length of LINEs ranged from 0.1 to 14.0 kb with a mean length of 1.3 kb, whereas the length of SINEs varied from 0.1 to 1.8 kb with a mean length of 1.5 kb. Similarly, a tremendous number of DNA transposons (24,386) belonging to class II-type TEs were scattered into foxtail millet genome. These transposons belong to various sub-classes, namely DNA/CMC-EnSpm, DNA/En-Spm, DNA/hAT-Ac, DNA/hAT-Tag1, DNA/hAT-Tip100, DNA/MULE-MuDR, DNA/PIF-Harbinger, DNA/TcMar-Stowaway, RC/Helitron, and DNA/Tourist based on similarity search with known TEs reported in other plant species (Fig. 3.4).

3.5.2 *Setaria viridis* (Green Millet)

Similar to *S. italica,* green millet contains a relatively lower amount of TEs. Comprehensive TE annotations of *S. viridis* revealed that ~40% TEs contributed to genome composition which includes both class I and II types of TEs. The whole genome sequence of green millet was scanned for representation of various types of TEs using DNA transposons and identified by RepeatMasker (http://www.repeatmasker.org/cgi-bin/WEBRepeatMasker). It was found that maximum number of Gypsy-type retrotransposons was 28,065 in the genome, whereas 14,253 Copia-type retrotransposons are present in all chromosomes of green millet (Fig. 3.5). Green millet was also curated for non-LTR elements and it was found that 3000 LINE elements are present in the genome. However, very low frequencies of SINEs were detected in all over the genome. DNA transposons (22,351) in green millet genome were also identified and it was found that a maximum number of DNA transposons belong to Harbinger category followed by DNA/CMC-EnSpm (6031) and DNA/TcMar-Stowaway (1571) (Fig. 3.5).

3.6 Epigenetic Regulation of TEs and Involvement in Gene Expression

TEs are known for jumping behavior which may produce more polymorphic locations in the genome, leading to the production of genetic and phenotypic variations among individuals. These phenomena could also create polymorphism

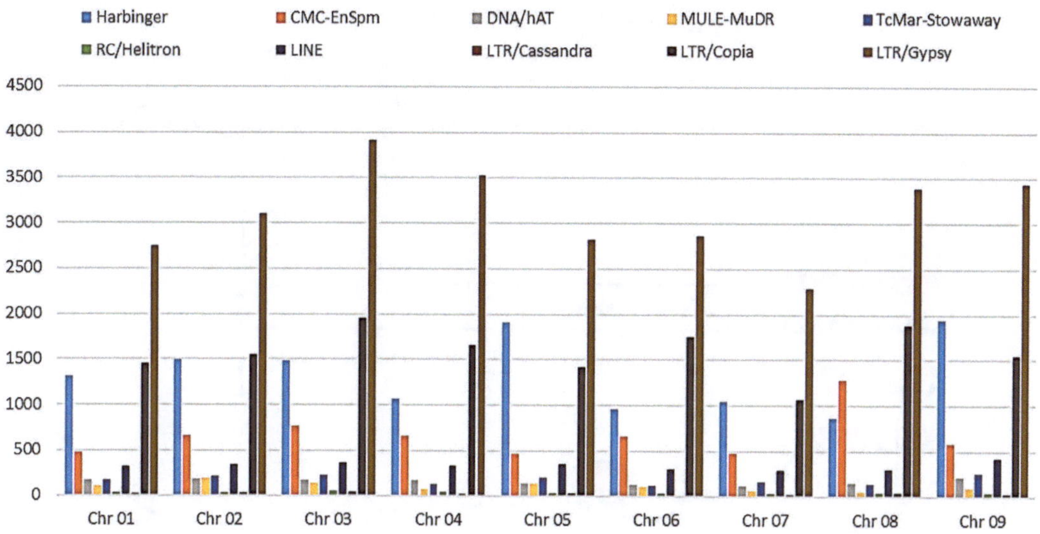

Fig. 3.5 Distribution of different classes of TEs among all nine chromosomes of *Setaria viridis*

within the cells or tissue of individuals by subjecting nearby genes to the epigenetic regulation that is targeted by TE. Various epigenetic mechanisms, such as the transcriptional and posttranscriptional mechanism of TE silencing, could involve regulation of TE translocation, which also affects the expression of nearby genes. DICER-LIKE1 (DCL1) is a member of the dicer family of proteins and is involved in one of the prime mechanism for epigenetic regulation of TE where dsRNA results from retrotransposons to form 21–30 nucleotides long small interfering RNAs (siRNA). These siRNAs are loaded onto RISC complex with the help of Argonaute protein for the degradation of the complimentary transcript (Fig. 3.6). Another important mechanism, DNA methylation, can bring changes to chromatin conformation that inhibits the binding of transcription factors, hence impeding transcription (Bell and Felsenfeld 2000). In higher eukaryotes, particularly in plants and fungi, DNA methylation is fundamentally described as a defense strategy that protects genomes against TEs by transcriptional silencing (Yoder et al. 1997; Zilberman et al. 2008). The first evidence for DNA methylation-mediated regulation of TE activity was reported in maize. Inactivation of the *Activator* (*Ac*), a TE, was correlated with the extent of DNA methylation in the transposase promoter sequences, and active *Ac* elements showed a reduced level of methylation in maize. (Chomet et al. 1987; Wang et al. 1996). Mutational studies in *Arabidopsis thaliana* recognized the reactivation of TEs in the *met1* and *ddm1* single mutants, which are defective in DNA-methylation (Lippman and Martienssen 2004; Miura et al. 2001).

DNA methylation alterations triggered in response to environmental stress can either get erased or persist in plant genome after the release of stress, imparting a stress memory. The stress memory is transmitted to progeny of the plant, preparing them for future stress encounters, and this is called transgenerational inheritance (Chinnusamy and Zhu 2009). In *Arabidopsis*, the RdDM mutants showed activated transposition of retrotransposon *ONSEN* under heat stress. This transposition was also observed in the progenies derived from tissues exposed to heat stress (Matsunaga et al. 2012). Further, preservation of altered DNA methylation level was observed in selfed progenies of rice genotypes exposed to salt and alkaline stress (Feng et al. 2012). Hence, inheritance of epigenetic modification elicited by stress can be anticipated as an approach to promote the development of more stress-tolerant crop varieties.

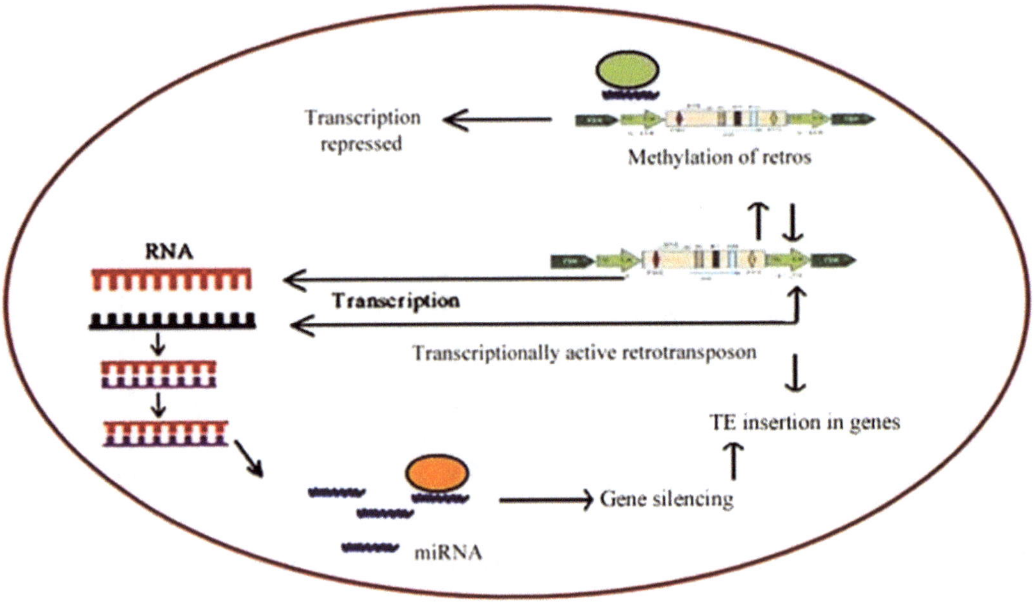

Fig. 3.6 Graphical representation on epigenetic regulation of TE and its impact on nearby genes influencing the gene expression

3.7 Utilization of TEs in Millets Improvements

Their ubiquity in plants, wide dispersion on all chromosomes, and activity in creating genomic diversity make TEs ideal for use as molecular markers (Schulman et al. 2004) which can also be used for genetic linkage mapping. Because of the rapid amplification activity of TE/retrotransposons through a "cut-and-paste" or a "copy-and-paste" mechanism, they are randomly inserted into other repetitive sequences, gene loci, or intergenic regions, resulting in different types of TE or repeat junctions. Most of those junctions are unique and genome-specific, and randomly distributed along chromosomes. In this regard, Devos et al. (2005) developed a technique where repeat junctions are used to generate unique markers. This technique can be useful for genome-wide marker development in large and repetitive genomes to differentiate the homologous region of millet genome. Several PCR-based molecular marker systems based upon TEs have been proposed and applied to genetic diversity assessment, and genetic and physical mapping, such as retrotransposon-based insertion polymorphism (Flavell et al. 1998), inter-retrotransposon amplified polymorphism (Kalendar and Schulman 2006), repeat junction marker (Wanjugi et al. 2009), repeat junction-junction marker (Luce et al. 2006), and insertion-site-based polymorphism (Devos et al. 2005; Paux et al. 2006). In foxtail millet, Yadav et al. (2015) successfully developed 20,278 PCR-based repeat junction markers using RJPrimers pipeline v1.0 (http://probes.pw.usda.gov/RJPrimers/). The complete foxtail genome was scanned to identify TE junction-based markers which are successfully validated in 99 foxtail millet lines for their utilization in diversity analysis. Retrotransposons-based markers for dissecting out the important QTL or genes could therefore be an important tool because they are ubiquitous in nature and represent the whole genome with high copy numbers. Another important property of retrotransposons is that they possess a replicative mode of transposition. Therefore the insertions are mostly stable. This is essential for determining parental lineage data in any study of phylogenetic relationships. Molecular marker technology based on SNPs and SSRs is reversible, which limits their use for such studies. Using various kind of technology using TE information could be utilized in exploring the phylogenetic relationships and biodiversity studies in millets. Moreover, retrotransposon-based markers also enable us to study the genetic linkage mapping, high-density mapping in target regions (Pearce et al. 2000).

3.8 Conclusions

As demonstrated above, the advent of next-generation sequencing has been a boon to whole genome sequencing and genome assembly analysis for their utilization in the genetic and epigenetic improvement of crop plants. With the availability of reference genome information, various bioinformatic tools have allowed us to understand the organization of TEs in the genome and its architecture. Being one of the major components of the plant genome, TE has an essential role in functional genome diversity and also impacts on phenotypic variations. The fractured and jumping nature of TEs has contributed to the heterochromatinization of the genome and evolution of linear chromosomes. The demonstration of genome architecture with respect to TEs in the genome could therefore provide the opportunity to decipher the role of TEs and nearby gene expression for regulating and shaping the crop phenotypic diversity. Nowadays, foxtail millet and green millet are being used as model crops for studying the genetics and genomics of millets, cereals, and bioenergy grasses. Hence, identifying TEs and classifying and analyzing the organization, and development of TE-based molecular markers in foxtail millet would serve as an important resource for millets, cereals, and bioenergy genomics.

Acknowledgements Studies on millet genomics in Dr. Manoj Prasad's laboratory are supported by the Science and Engineering Research Board (SERB), Department of Science and Technology (DST), Government of India [Grant No. EMR/2015/000464], by the Department of Biotechnology, Government of India [Grant No. BT/HRD/NBA/37/01/2014], and by Core Grant of National Institute of Plant Genome Research (NIPGR), New Delhi, India. Dr. Chandra Bhan Yadav acknowledges the award of Young Scientist Research Grant from DST-SERB, Government of India [File No. YSS/2015/000287].

References

Allen T, Koustenis A, Theodorou G, Somers DE, Kay SA, Whitelam GC, Devlin PF (2006) *Arabidopsis* FHY3 specifically gates phytochrome signaling to the circadian clock. Plant Cell 18:2506

Bandyopadhyay T, Muthamilarasan M, Prasad M (2017) Millets for Next Generation Climate-Smart Agriculture. Front Plant Sci 8:1266

Bell AC, Felsenfeld G (2000) Methylation of a CTCF-dependent boundary controls imprinted expression of the Igf2 gene. Nature 405:482–485

Bennetzen JE, Kellogg A (1997) Do plants have a one-way ticket to genomic obesity? Plant Cell 9:1509–1514

Bennetzen JL, Schmutz J, Wang H, Percifield R, Hawkins J, Pontaroli AC, Estep M, Feng L, Vaughn JN, Grimwood J, Jenkins J, Barry K, Lindquist E, Hellsten U, Deshpande S, Wang X, Wu X, Mitros T, Triplett J, Yang X, Ye CY, Mauro-Herrera M, Wang L, Li P, Sharma M, Sharma R, Ronald PC, Panaud O, Kellogg EA, Brutnell TP, Doust AN, Tuskan GA, Rokhsar D, Devos KM (2012) Reference genome sequence of the model plant Setaria. Nature Biotechnol 30:555–561

Brenchley R, Spannagl M, Pfeifer M, Barker GL, D'Amore R, Allen AM, McKenzie N, Kramer M, Kerhornou A, Bolser D, Kay S, Waite D, Trick M, Bancroft I, Gu Y, Huo N, Luo MC, Sehgal S, Gill B, Kianian S, Anderson O, Kersey P, Dvorak J, McCombie WR, Hall A, Mayer KF, Edwards KJ, Bevan MW, Hall N (2012) Analysis of the bread wheat genome using whole-genome shotgun sequencing. Nature 491:705–710

Chinnusamy V, Zhu JK (2009) Epigenetic regulation of stress responses in plants. Curr Opin Plant Biol 12 (2):133–139

Chomet PS, Wessler S, Dellaporta SL (1987) Inactivation of the maize transposable element Activator (Ac) is associated with its DNA modification. EMBO J 6:295–302

Devos KM, Ma J, Pontaroli AC, Pratt LH, Bennetzen JL (2005) Analysis and mapping of randomly chosen bacterial artificial chromosome clones from hexaploid bread wheat. Proc Natl Acad Sci USA 102:19243–19248

Durán-Figueroa N, Vielle-Calzada JP (2010) ARGONAUTE9-dependent silencing of transposable elements in pericentromeric regions of *Arabidopsis*. Plant Signal Behav 5(11):1476–1479

Dwivedi S, Upadhyaya H, Senthilvel S, Hash C, Fukunaga K, Diao X, Santra D, Altensperger D, Prasad M (2012) Millets: genetic and genomic resources. In:

Janick J (ed) Plant Breed Rev, vol 35. John Wiley & Sons Inc, USA, pp 247–375
Echenique V, Stamova B, Wolters P, Lazo G, Carollo L, Dubcovsky J (2002) Frequencies of Ty1-copia and Ty3- gypsy retroelements within the Triticeae EST databases. Theor Appl Genet 104:840–844
FAO (1995) Sorghum and millets in human nutrition. FAO food and nutrition series No 27. Rome, Italy: Food and Agriculture Organization, 184
Feng Q, Yang C, Lin X, Wang J, Ou X, Zhang C, Chen Y, Liu B (2012) Salt and alkaline stress induced transgenerational alteration in DNA methylation of rice (Oryza sativa). AJCS 6(5):877–883
Feschotte C, Jiang N, Wessler SR (2002) Plant transposable elements: where genetics meets genomics. Nat Rev Genet 3:329–341
Flavell AJ, Knox MR, Pearce SR, Ellis TH (1998) Retrotransposon-based insertion polymorphisms (RBIP) for high throughput marker analysis. Plant J 16:643–650
Ghannoum O, Evans JR, von Caemmerer S (2011) Nitrogen and water use efficiency in C4 plants. In: Raghavendra AS, Sage RF (eds) C4 photosynthesis and related CO2 concentrating mechanisms. Springer, The Netherlands, pp 129–146
Gupta S, Kumari K, Das J, Lata C, Puranik S, Prasad M (2011) Development and utilization of novel intron length polymorphic markers in foxtail millet [*Setaria italica* (L.) P. Beauv.]. Genome 54:586–602
Gupta S, Kumari K, Sahu PP, Vidapu S, Prasad M (2012) Sequence based novel genomic microsatellite markers for robust genotyping purposes in foxtail millet [*Setaria italica* (L.) P. Beauv.]. Plant Cell Rep 31:323–337
Jia G, Huang X, Zhi H, Zhao Y, Zhao Q, Li W et al (2013) A haplotype map of genomic variations and genome-wide association studies of agronomic traits in foxtail millet (*Setaria italica*). Nat Genet 45 (8):957–961
Kalendar R, Schulman AH (2006) IRAP and REMAP for retrotransposon-based genotyping and fingerprinting. Nat Protoc 1:2478–2484
Kocacinar F, Mckown AD, Sage TL, Sage RF (2008) Photosynthetic pathway influences xylem structure and function in *Flaveria* (*Asteraceae*). Plant, Cell Environ 31:1363–1376
Kumar A, Bennetzen JL (1999) Plant retrotransposons. Annu Rev Genet 33:479–532
Lippman Z, Martienssen R (2004) The role of RNA interference in heterochromatic silencing. Nature 431 (7006):364–370
Lisch D (2009) Epigenetic regulation of transposable elements in plants. Annu Rev Plant Biol 60:43
Lisch D, Bennetzen JL (2011) Transposable element origins of epigenetic gene regulation. Curr Opin Plant Biol 14:156
Long SP (1999) Environmental responses. In: Sage RF, Monson RK (eds) C4 plant biology. Academic Press, San Diego, pp 215–249
Luce AC, Sharma A, Mollere OS, Wolfgruber TK, Nagaki K, Jiang J, Presting GG, Dawe RK (2006) Precise centromere mapping using a combination of repeat junction markers and chromatin immunoprecipitation-polymerase chain reaction. Genetics 174:1057–1061
Macas J, Neumann P, Pozárková D (2003) Zaba: a novel miniature transposable element present in genomes of legume plants. Mol Genet Genomics 269:624–631
Matsunaga W, Kobayashi A, Kato A, Ito H (2012) The effects of heat induction and the siRNA biogenesis pathway on the transgenerational transposition of ONSEN, a copia-like retrotransposon in *Arabidopsis thaliana*. Plant Cell Physiol 53:824–833
Meyers BC, Tingey SV, Morgante M (2001) Abundance, distribution, and transcriptional activity of repetitive elements in the maize genome. Genome Res 11:1660–1676
Miura A, Yonebayashi S, Watanabe K, Toyama T, Shimada H, Kakutani T (2001) Mobilization of transposons by a mutation abolishing full DNA methylation in *Arabidopsis*. Nature 411:212–214
Muthamilarasan M, Prasad M (2015) Advances in Setaria genomics for genetic improvement of cereals and bioenergy grasses. Theor Appl Genet 128:1–14
Muthamilarasan M, Venkata Suresh B, Pandey G, Kumari K, Parida SK, Prasad M (2014) Development of 5123 intron-length polymorphic markers for large-scale genotyping applications in foxtail millet. DNA Res 21:41–52
Muthamilarasan M, Dhaka A, Yadav R, Prasad M (2016) Exploration of millet models for developing nutrient rich graminaceous crops. Plant Sci 242:89–97
Olmedo-Monfil V, Durán-Figueroa N, Arteaga-Vázquez M, Demesa-Arévalo E, Autran D, Grimanelli D, Slotkin RK, Martienssen RA, Vielle-Calzada J-P (2010) Control of female gamete formation by a small RNA pathway in *Arabidopsis*. Nature 464:628–632
Pandey G, Misra G, Kumari K, Gupta S, Parida SK, Chattopadhyay D, Prasad M (2013) Genome-wide development and use of microsatellite markers for large-scale genotyping applications in foxtail millet [*Setaria italica* (L.)]. DNA Res 20:197–207
Paux E, Roger D, Badaeva E, Gay G, Bernard M, Sourdille P, Feuillet C (2006) Characterizing the composition and evolution of homoeologous genomes in hexaploid wheat through BAC-end sequencing on chromosome 3B. Plant J 48:463–474
Pearce SR et al (2000) Pea Ty1-copia group retrotransposons: transpositional activity and use as markers to study genetic diversity in Pisum. Mol Gen Genet 263:898–907
Peaston AE, Evsikov AV, Graber JH, de Vries WN, Holbrook AE, Solter D, Knowles BB (2004) Retrotransposons regulate host genes in mouse oocytes and preimplantation embryos. Dev Cell 7:597–606
Qie L, Jia G, Zhang W, Schnable J, Shang Z, Li W et al (2014) Mapping of quantitative trait locus (QTLs) that contribute to germination and early seedling drought

tolerance in the interspecific cross Setaria italic x *Setaria viridis*. PLoS ONE 9(7):e101868

Rabinowicz PD, Schutz K, Dedhia N, Yordan C, Parnell LD, Stein L, McCombei WR, Martienssen RA (1999) Differential methylation of genes and retrotransposons facilitates shotgun sequencing of the maize genome. Nat Genet 23:305–308

Sabot F, Simon D, Bernard M (2004) Plant transposable elements, with an emphasis on grass species. Euphytica 139:227–247

Sage RF, Pearcy RW (2000) The physiological ecology of C4 photosynthesis. In: Leegood RC, Sharkey TD, von Caemmerer S (eds) Photosynthesis: physiology and metabolism. Kluwer Academic, Dordrecht, The Netherlands, pp 497–532

SanMiguel P, Tikhonov A, Jin Y-K, Motchoulskaia N, Zakharov D, Melake-Berhan A, Springer PS, Edwards KJ, Lee M, Avramova Z, Bennetzen JL (1996) Nested retrotransposons in the intergenic regions of the maize genome. Science 274:765–768

SanMiguel P, Gaut BS, Tikhonov A, Nakajima Y, Bennetzen JL (1998) The paleontology of intergene retrotransposons of maize. Nat Genet 20:43–45

Schulman AH, Flavell AJ, Ellis THN (2004) The Application of LTR Retrotransposons as Molecular Markers in Plants. Methods Mol Biol 859:115–153

Tang W et al (2012) Transposase-derived proteins FHY3/FAR1 interact with PHYTOCHROME-INTERACTING FACTOR1 to regulate chlorophyll biosynthesis by modulating HEMB1 during deetiolation in *Arabidopsis*. Plant Cell 24:1984

The Arabidopsis Genome Initiative (2000) Analysis of the genome sequence of the flowering plant *Arabidopsisthaliana*. Nature 408:796–815

The Rice Chromosome 10 Sequencing Consortium (2003) In-Depth view of structure, activity, and evolution of rice chromosome 10. Science 300:1566–1569

Vicient CM (2010) Transcriptional activity of transposable elements in maize. BMC Genom 11:601

Wang L, Heinlein M, Kunze R (1996) Methylation pattern of activator (Ac) transposase binding sites in maize endosperm. Cell 8:747–758

Wanjugi H, Coleman-Derr D, Huo N, Kianian SF, Luo MC, Wu J, Anderson O, Gu YQ (2009) Rapid development of PCR-based genome-specific repetitive DNA junction markers in wheat. Genome 52:576–587

Yadav CB, Suresh BV, Muthamilarasan M, Pandey G, Khan Y, Prasad M (2015) Genome-wide development of transposable elements-based markers in foxtail millet (*Setaria italica* L.) and construction of an integrated database. DNA Res 22(1):79–90

Yoder JA, Walsh CP, Bestor TH (1997) Cytosine methylation and the ecology of intragenomic parasites. Trends Genet 13(8):335–340

Zilberman D, Coleman-Derr D, Ballinger T, Henikoff S (2008) Histone H2A.Z and DNA methylation are mutually antagonistic chromatin marks. Nature 456 (7218):125–129

Exploiting Genome Sequence Information to Develop Genomic Resources for Foxtail Millet Improvement

4

Mehanathan Muthamilarasan and Manoj Prasad

Abstract

Foxtail millet is an excellent model crop for studying the biology of C_4 photosynthesis, abiotic stress tolerance, and biofuel traits. In addition, grains of foxtail millet are rich in proteins, micro- and macro-nutrients, and other bioactive compounds. Being an old domesticated crop, foxtail millet has contributed to the development of human civilization and is still grown as a staple food, particularly in India and China. India is the largest producer of millets. However, the breeding technology used for foxtail millet is far behind that of major millets such as pearl millet and finger millet, and major cereals such as rice and wheat. Development of genomic resources is the first step toward improving the breeding strategies, which subsequently lead to the development of elite cultivars with higher yield and desirable agronomic traits. Much national and international effort has been invested in this regard to develop genetic and genomic resources and these have produced significant outcomes. In this context, the present chapter enumerates the genetic and genomic resources available for foxtail millet improvement. In particular, the chapter discusses the development of novel molecular markers, their application in genomics-assisted breeding, and the construction of integrated databases. In addition, the genes and families identified in foxtail millet which have relevance to growth, development, and stress response have also been summarized.

4.1 Introduction

Advances in next-generation sequencing technology and the introduction of high-throughput genome analysis platforms have accelerated the development of tools and strategies useful for plant breeding. Genome-wide association studies (GWAS) facilitate the identification of genomic

M. Muthamilarasan · M. Prasad (✉)
National Institute of Plant Genome Research (NIPGR), Aruna Asaf Ali Marg, New Delhi 110067, India
e-mail: manoj_prasad@nipgr.ac.in

M. Prasad (ed.), *The Foxtail Millet Genome*, Compendium of Plant Genomes,
DOI 10.1007/978-3-319-65617-5_4

regions regulating phenotypic variations in a set of germplasm (Brachi et al. 2011; Hamblin et al. 2011); whereas, genomic selection (GS) allows the selection of superior genotypes based on genomic estimated breeding values estimated from molecular marker data (Lorenz et al. 2011; Kole et al. 2015). Being evolved from traditional biparental quantitative trait locus (QTL) mapping and marker-assisted selection, GWAS and GS approaches possess broad-spectrum applications including the potential to expedite genomics-assisted breeding (GAB). Primarily, GAB relies on the development of genetic and genomic resources which are further deployed to generate novel lines with improved agronomic traits. Genetic resources denote the germplasm collection which includes the existing diversity, base collection, core collection, core reference set, nucleus sample, and panel. Genomic resources include the molecular markers, genome and transcriptome assemblies, genetic and physical maps, and QTLs as well as candidate genes. GAB for crop improvement has been successfully demonstrated in several crops including legumes (Pazhamala et al. 2015), fruit trees (Iwata et al. 2016), and perennial crops (Migicovsky and Myles 2017). However, in millets, the application of GAB is limited. In a functional genomics perspective, several genes and corresponding gene families have been characterized in foxtail millet and, interestingly, the number of gene families studied for their relevance in the improvement of foxtail millet is higher than other millets (Muthamilarasan and Prasad 2015). The availability of draft genome sequence data in the public database has facilitated these studies at a genome-wide level, which has identified several candidate genes involved in abiotic stress response.

Millets are small-grained panicoid crops possessing C_4 photosynthetic traits and better water-use and nitrogen-use efficiencies (Muthamilarasan et al. 2016a). Among millets, foxtail millet (*Setaria italica* L.) is considered to be the model crop for studying the systems biology of other millets, biofuel crops, and cereals (Lata et al. 2013; Muthamilarasan and Prasad 2015). Being a C_4 crop, foxtail millet is naturally equipped with better water-use and nitrogen-use efficiencies, anda few morpho-physiological traits, including dense and deep root systems, smaller leaf area, and thickening of cell walls, were suggested to confer durable tolerance to broad-spectrum abiotic stresses (Lata et al. 2013; Diao et al. 2014). Furthermore, the seeds of foxtail millet require only 26% of their seed weight in water to germinate, whereas other cereals require a minimum of 45% of their seed weight. Similarly, foxtail millet requires only 257 g of water to produce 1 g dry biomass, which is the lowest among cereals, as maize and wheat require 470 and 510 g, respectively (Diao et al. 2014). This highlights the climate resilient characteristics of foxtail millet. Further studies on genetics and genomics of stress biology have shown that the crop has novel as well as known stress-responsive genes which participate in stress adaptation (Muthamilarasan and Prasad 2015). Among abiotic stresses, drought and salinity are the immediate impacts of climate change, and in this scenario the global research community is actively involved in understanding the genetics and genomics of crops tolerant to these stresses, with the aim of engineering these traits in susceptible plants (Kole et al. 2015).

Foxtail millet is extensively cultivated in arid and semi-arid regions of the world, with China and India being the largest producers (Bandyopadhyay et al. 2017). However, its grain productivity and yield potential are still low compared to other major cereals. Because of the lack of an efficient transformation system (discussed in Chap. 7), genetic improvement of foxtail millet through transgene-based approaches is not feasible, and therefore GAB appears promising in establishing a modern breeding forum for genomic designing of elite foxtail millet cultivars. Toward achieving this, efforts have been made to collect and conserve the germplasm resources and to develop several high-throughput, genome-wide molecular markers. Foxtail millet genomics has also seen the development of several web-based databases to bring the developed resources to the global research community. Given this, the present chapter summarizes the progress and prospects of developing genetic and genomic resources for the improvement of foxtail millet.

4.2 Genetic Resources Available for Crop Improvement

Germplasm collection is the foremost resource for selection and breeding of elite lines with improved agronomic traits. These germplasm collections include the existing diversity, base collection, core collection, core reference set, nucleus sample, and panel. The core collection includes the representative samples from existing diversity and base collection, whereas the core reference set indicates the preferred material for multiple characterizations and data integration. Similarly, nucleus sample indicates the germplasm recommended for inclusion in any characterization studies, whereas panel denotes the germplasm material adapted to a particular experiment with a specific purpose. In this context, not many resources are available for all these different experimental representative sets; however, the International Crop Research Institute for the Semi-Arid Tropics (ICRISAT), India is holding 1474 cultivated germplasm accessions from 23 countries (Upadhyaya et al. 2008). From this, a core collection of 155 accessions has been obtained for application in association mapping studies (Upadhyaya et al. 2008). Similarly, National Institute of Plant Genome Research, India possesses around 200 accessions collected from diverse eco-geographical regions of the world, and this association panel has proved to be useful in GAB (Gupta et al. 2014). In addition, National Bureau of Plant Genetic Resources (NBPGR), and All India Coordinated Small Millets Improvement Project (AICSMIP), India, possess hundreds of foxtail millet accessions, and it should be noted that these institutes have several other species of *Setaria* including *S. viridis*, *S. parviflora*, *S. sphacelata*, *S. verticillata*, *S. sphacelata (subsp. anceps)*, *S. neglecta*, *S. lachnea*, *S. incrassate*, *S. australiensis*, *S. chevalieri*, and *S. pumila* (Saha et al. 2016). Furthermore, several international organizations also maintain the germplasm of foxtail millet. These include Institute of Crop Sciences Germplasm Resources (China), Plant Genetic Resources Conservation Unit (USDA, USA), the N. I. Vavilov Research Institute of Plant Industry (Russia), Centre de coopération internationale en recherche agronomique pour le développement (France), National Institute of Agrobiological Sciences (Japan), Kenya Agricultural Research Institute (Kenya), and Institute of Biodiversity Conservation, (Ethiopia). These collections assist in establishing new collections such as a core collection, core reference set, nucleus sample, and panel for their further downstream utilization in crop improvement.

4.3 Genomic Resources Developed in Foxtail Millet

Foxtail millet has illustrated a leap in the development of genomics after the release of genome sequence information. In view of this, the following section provides an overview of foxtail millet genome followed by the availability of molecular markers and the details of genomic regions governing agronomic traits.

4.3.1 Genome Sequence Information

The genome of foxtail millet has been sequenced by Beijing Genome Initiative (BGI; Zhang et al. 2012), China and the United States Department of Energy-Joint Genome Initiative (USDOE-JGI), USA (Bennetzen et al. 2012). BGI has sequenced the cultivar 'Zhang gu' using Illumina second generation sequencers. The project generated 16,903 contigs and 439 scaffolds representing 423 Mb of the predicted genome size of 485 Mb. A photo-thermosensitive male-sterile line 'A2' has also been sequenced, and the alignment of the reads with cv. 'Zhang gu' generated millions of single nucleotide polymorphisms (SNPs) and insertion-deletions (InDels) which could be exploited as functional markers (FMs). Using this information, a genetic linkage map was constructed from an F_2 mapping population of $\sim$500 accessions developed from a cross made between 'Zhang gu' and 'A2' (Zhang et al. 2012). Similarly, the USDOE-JGI sequenced an inbred line 'Yugu1' along with a wild (*S. viridis*, green foxtail) accession A10

(Bennetzen et al. 2012). The sequencing was performed using ABI3730xl capillary sequencer which generated ~400 Mb data. A recombinant inbred line population consisting of 247 progenies developed from a cross between foxtail millet accession 'B100' and green foxtail 'A10' was sequenced to construct a genetic map consisting of 9 linkage groups using 992 SNPs (Bennetzen et al. 2012). The release of genome sequence information has expedited the structural and functional genomics studies in foxtail millet along with the development of genomic resources for this important crop. In addition, Zhang et al. (2012) have studied the organization and evolution of C_4 pathway genes in foxtail millet in comparison to C_3 grasses. The study showed that the genes involved in C_4 carbon fixation pathways were also present in C_3 plants, and it was therefore inferred that the evolution of the C_4 pathway might have resulted from functional changes in these genes. The study also identified that the gene *Ft_CA1* might play an important role in the C_4 pathway of foxtail millet (Zhang et al. 2012). This suggests that foxtail millet could be an emerging model for dissecting C_4 genetics and physiology.

4.3.2 Molecular Markers Developed in Foxtail Millet

Development of genomic resources including large-scale, genome-wide molecular markers is the first step toward improving the breeding strategies, which subsequently lead to the development of elite cultivars with desirable agronomic traits. Molecular markers are DNA fragments/sequences used to spot genetic polymorphism within a population or between individuals. The basis for polymorphism is base pair deletion, substitution, addition, and variation in sequence length. These markers are broadly classified into genomic and genic markers, where the latter are present within a gene and the former is present in the regions other than a gene. The genic markers are of more importance than the genomic markers as they are developed directly from the gene sequence, which could be in the form of bacterial artificial chromosomes, cDNA libraries, ESTs (expressed sequence tags), and transcriptome sequences (Varshney et al. 2007). Based on the site of polymorphism and their subsequent effect on phenotype, genic markers are categorized into two types, namely gene-targeted markers (GTM) and FMs (Andersen and Lubberstedt 2003). GTMs are polymorphisms derived from the genes which do not have any effect on phenotypic variation (Schmitt et al. 2006; Aggarwal et al. 2007), whereas FMs do affect the phenotypic trait variation (Varshney et al. 2007). For example, the candidate gene-based molecular markers have their effect on phenotype and based on their role in phenotypic variation; these markers are further classified into two sub-groups, namely indirect and direct FMs (Varshney et al. 2007). A schematic representation of the development of genic markers is shown in Fig. 4.1. For this, the sequence data of genes/ESTs are processed in two different ways, namely direct mapping and in silico mining. In direct mapping, the cDNA clones of ESTs are used as restriction fragment length polymorphism (RFLP) probes or primers for sequence tagged sites or cleaved amplified polymorphic sequence-based markers are designed for each EST. In the case of in silico mining, software for identification of simple sequence repeats (SSRs) and/or SNPs is used, which also designs the primers flanking these markers. However, for the development of FMs, a set of functionally characterized genes, their allele sequences, along with polymorphic, functional motifs affecting the phenotype within these genes and the association of the polymorphism with trait variation are required (Varshney et al. 2007).

The first markers in foxtail millet were RFLP markers reported by Wang et al. (1998). An RFLP-based map consisting of 160 loci was constructed in an intervarietal cross, and these markers proved useful in constructing comparative genetic maps between foxtail millet and rice genomes (Devos et al. 1998). Following this, Jia et al. (2007) developed 26 EST-derived SSRs and the same group has constructed an SSR-linkage map by integrating 81 SSR markers [enriched for (GA)n and (CA)n] with 20 RFLP anchored markers (Jia et al. 2009). Later, Gupta et al. (2011) developed 98 intron length polymorphic (ILP) markers by exploiting the EST data of dehydration- and salinity-stressed suppression subtractive hybridization libraries constructed by Lata et al. (2010), Puranik et al. (2011a, b), respectively. Similarly, microsatellite enriched libraries were constructed for the $(CA)_n$, $(AAC)_n$, and $(ATG)_n$ sequences (Gupta et al. 2012a, b) and $(GA/CT)_n$ sequence to develop 172 and 78 markers, respectively (Gupta et al. 2013). However, the area of structural genomics has shown a leap after the release of genome sequence information (Fig. 4.2). There were several genome-wide markers developed in large-scale and their utility in genotyping purposes was demonstrated (Table 4.1). The following sections elaborate the different markers developed post-genome sequence era of foxtail millet.

4.3.2.1 Genomic Microsatellite Markers

Next-generation sequencing technologies have generated enormous datasets of genomic

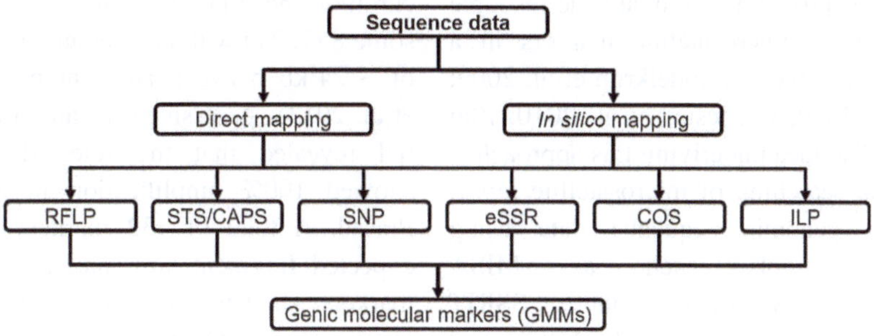

Fig. 4.1 Flowchart showing the development of genic molecular markers. RFLP—Restriction fragment length polymorphism; STS—sequence tagged sites; CAPS—cleaved amplified polymorphic sequence; SNP—single nucleotide polymorphism; SSR—simple sequence repeat; COS—conserved orthologous sequence; ILP—intron length polymorphism

Fig. 4.2 General strategy used for development of genome-wide molecular markers to demonstrate their application in large-scale genotyping purposes in foxtail millet. NGS—Next generation sequencing; EST—expressed sequence tags; MISA—MIcroSAtellite identification tool; TE—transposable elements; SSR—simple sequence repeats; ILP—intron length polymorphism

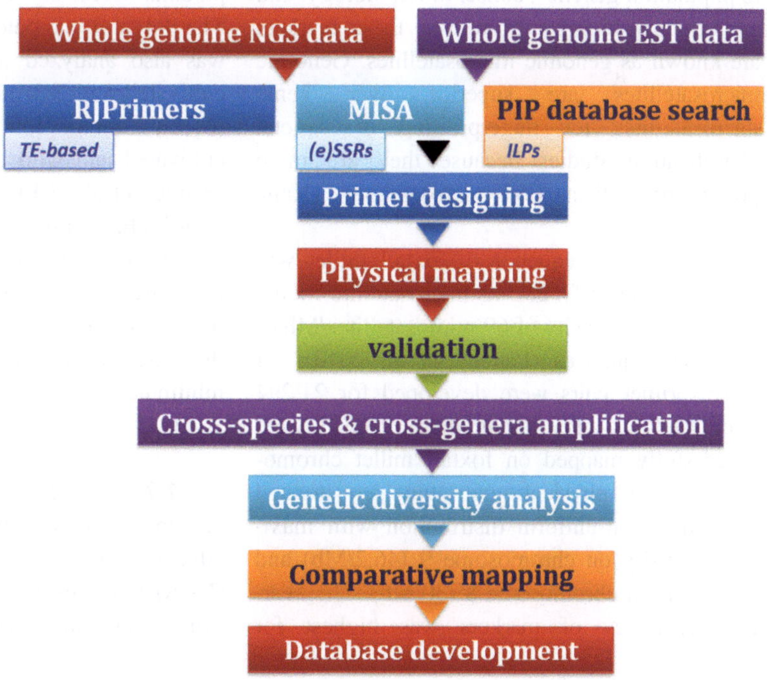

Table 4.1 Details of large-scale molecular markers developed in foxtail millet post-genome sequence release

Type of marker	Number of markers developed	Percentage polymorphism (%)	Percentage cross-transferability	Genetic diversity range	Reference
SSR	21,294	~40	~89%	Not analyzed	Pandey et al. (2013)
EST-SSR	534	~60	~88%	0.02–0.65	Kumari et al. (2013)
ILP	5123	~45	~85%	0.03–0.52	Muthamilarasan et al. (2014a, b, c)
miRNA-based	176	~53	~70%	Not analyzed	Yadav et al. (2014)
TE-based	20.278	~40	Not analyzed	0.03–0.98	Yadav et al. (2015a, b)

sequences, thus providing an invaluable resource for generation of microsatellite markers in a number of plant species (Abdelkrim et al. 2009; Castoe et al. 2010; Csencsics et al. 2010; Zhu et al. 2012). The idea underlying this approach is computational searching of microsatellite repeat motifs from genomic sequence data using bioinformatics tools such as MISA (MIcroSAtellite), SSR finder, Sputnik, SSRIT (SSR Identification Tool), SSR SEARCH and TRF (Tandem Repeat Finder). Primers are designed from the adjacent regions of repeats and then developed as microsatellite markers. These markers have wide applicability in studies related to population genetics, conservation biology, and evolutionary biology. Moreover, these markers are known as genomic microsatellites. Genomic microsatellites are superior over genic microsatellites for fingerprinting or varietal identification studies because they are more polymorphic than EST-microsatellites (Kalia et al. 2011).

The genome sequence of foxtail millet was scanned to identify 28,342 microsatellite motifs at an average density of 69 microsatellites/Mb of genomic sequences (Pandey et al. 2013). Of these, primer pairs were developed for 21,294 microsatellites and, among these, 15,573 were successfully mapped on foxtail millet chromosomes. The physical map thus constructed showed a non-uniform distribution with maximum density on chromosome 9 (46.2/Mb) and minimum on chromosome 8 (30/Mb). The average frequency of markers was highest for chromosome 9 (17.5%) and lowest for chromosome 8 (7.8%) with an average physical gap size of ~24 kb between adjacent markers (Pandey et al. 2013). Assessment of amplification potential revealed that the selected 159 markers showed 100% amplification in *S. italica* cv. Prasad. A total of 132 markers produced the expected fragment size, and the remaining 27 markers showed deviation from the desired amplicon size. Of 159 markers, 152 amplified single alleles while the remaining 7 amplified multiple alleles. Similarly, polymorphism assay showed that ~19% of 159 markers produced polymorphism between Prasad and Lepakshi, the parental accession of an F_2 mapping population. The polymorphic potential of developed markers was also analyzed in 8 *Setaria* accessions (4 cultivated and 3 wild) and it revealed that 107 (67.2%) out of 159 showed polymorphism in cultivated and wild accessions of foxtail millet (Pandey et al. 2013). Furthermore, cross genera transferability assay among millets, cereals, and bioenergy grass species showed average transferability of 89% for these markers. The maximum transferability percentage was observed in the case of guinea grass, that is, 98.2% and minimum in wheat, that is, 71.2% (Pandey et al. 2013).

4.3.2.2 Genic Microsatellite Markers

Advances in sequencing technology and the advent of next generation sequencing (NGS) have resulted in several gene discovery projects in plant species, which ultimately led to accumulation of the information in the public domain. NCBI dbEST (https://www.ncbi.nlm.nih.gov/genbank/dbest/dbest_access/), a division of GenBank, serves as the single standalone database for the ESTs (Boguski et al. 1993). These ESTs are retrievable and searchable for the presence of SSR motifs. These SSRs are called EST-derived SSRs (eSSRs) or genic microsatellites. Several bioinformatics tools are available for searching the SSR motifs in ESTs, and for designing primers flanking the SSRs. Popular software includes MISA (MIcroSAtellite), SSRfinder, Sputnik, SSRIT (SSR Identification Tool), SSRSEARCH, and TRF (Tandem Repeat Finder) (Kalia et al. 2011). The unrestricted access to EST data in the public domain facilitates easy accessibility to these data and development of eSSRs using the open-resource software. The greatest advantage of genic SSRs over genomic SSRs is that the former detect sequence variations in the genes and therefore enable a perfect marker-trait association. In addition, these markers would have higher transferability percentage across related genera and species because of the conservedness of genes among the genomes (Gupta and Varshney 2000). However, higher sequence conservation in transcribed regions results in lower polymorphism of these eSSRs, which in turn affects the efficacy of the markers in genetic diversity analyses (Kalia et al. 2011). In foxtail millet, 66,027 ESTs reported in NCBI dbEST were assembled to generate 24,828 unigenes, from which 534 eSSRs were developed (Kumari et al. 2013). Furthermore, these markers were used for constructing the physical map, and their utility was demonstrated in gene mapping, germplasm characterization, and analysis of genetic diversity and genetic relationships (Kumari et al. 2013).

4.3.2.3 ILP Markers

During the course of evolution, introns were under low purifying selection pressure, and therefore they are less conserved but highly variable (Badoni et al. 2016). These polymorphisms within the introns can be exploited to develop markers, among which, ILP markers are easily identifiable (Wang et al. 2005). The introns encompassing sequence polymorphism are detected, and primers are designed from the flanking exons to enable amplification through exon-primed intron-crossing PCR (EPIC-PCR; Palumbi 1995). Because of the conserved nature of exons, the primers designed from the exons have more widespread applications than the primers designed from non-coding regions (Wang et al. 2005). The first study on the development of ILPs and their application in genotyping was reported in rice by Wang et al. (2006), following which the markers were developed in several other crops, including *Medicago* (Choi et al. 2004), *Brassica* (Panjabi et al. 2008), wheat (Zhou et al. 2010), soybean (Shu et al. 2010), tomato (Wang et al. 2010), chickpea (Choudhary et al. 2012), cowpea (Gupta et al. 2012a, b), maize (He et al. 2015), and sorghum (Jaikishan et al. 2015). In foxtail millet, Muthamilarasan et al. (2014a, b, c) have exploited the EST database to develop 5123 ILP markers. Furthermore, the group has demonstrated the applicability of these markers in genetic diversity analysis, construction of high-density genetic linkage maps, comparative genome mapping, evolutionary studies, mapping of genes/QTLs regulating important agronomic traits, and marker-assisted breeding (Muthamilarasan et al. 2014a; Muthamilarasan and Prasad 2015).

4.3.2.4 MicroRNA-Based Markers

MicroRNAs (miRNAs) are ~18- to 24-nucleotide long non-coding RNAs (ribonucleic acids), which regulate the gene expression at transcriptional and post-transcriptional levels by cleavage or translational inhibition of target mRNA (reviewed in Muthamilarasan and Prasad 2013). The SSR signatures in these miRNAs were identified to develop molecular markers, and these markers proved successful in animal systems (Grady and Tewari 2010; Fu et al. 2011). The advantages of these markers, including high reproducibility and high polymorphism, have helped their development in plant systems. For this, pre-miRNA sequences were searched against the genome sequence of respective crops, and 500 bp of flanking sequences were retrieved from the genomic regions showing, perfect alignment (without mismatch on mature sequences). The retrieved sequences (1000 bp along with pre-miRNA sequence) were searched for the presence of SSRs using in silico identification tools, and the primers were designed flanking these SSRs (Mondal and Ganie 2014). In an alternative method, pre-miRNA sequences from the target species along with other related species were retrieved (from the public domain, such as miRBase). The sequences were then aligned to identify the conserved regions, and primers were designed for these conserved regions corresponding to stem-loops of pre-miRNA sequences (Fu et al. 2013). Fu et al. (2013) were the first to report miRNA-based markers in *Brassica* species using method II. They developed 46 miRNA-based markers, of which 28% were polymorphic, and these markers exhibited significant transferability across *Brassica* species. Mondal and Ganie (2014) were the first to demonstrate method I for developing miRNA-based SSR markers in rice. The study developed 12 markers, which were capable of differentiating salt tolerant and susceptible genotypes in rice. Similarly, the authors had also performed a genome-wide analysis of miRNA-based SSRs in rice to develop 129 markers (Ganie and Mondal 2015). In view of this, Yadav et al. (2014) have analyzed the genome-wide miRNAs reported in foxtail millet and other related species to develop 176 miRNA-based markers. Further utility of these markers in genotyping purposes demonstrated that miRNA-based markers exhibit high polymorphism, efficiency, reproducibility, stability, and cross-transferability (Yadav et al. 2014).

4.3.2.5 Transposable Elements-Based Markers

A significant fraction (more than 50%) of plant genome is constituted by transposable elements (TEs), which are capable of changing their position in the genome through transposition (Grzebelus 2006). This phenomenon acted as a major driving force during the course of genome evolution. These TEs are classified into class I and II, based on their mode of replication and transposition. The genome-wide distribution of TEs, along with the abundance, availability, and variability of closely-related genomes, have enabled the development of TE-based molecular markers. Most of these markers utilize the insertion sites of TEs, as TE insertion produces unique junctions having boundaries between the transposon and the DNA sequence at the site of insertion (Bennetzen 2000; Kalendar et al. 2011). Recently, the RJPrimers tool has been established for developing TE junction-based markers (You et al. 2010), which could be of five classes, namely repeat junction markers (RJM; Wanjugi et al. 2009), repeat junction-junction markers (RJJM; Luce et al. 2006), insertion-site-based polymorphism (ISBP; Devos et al. 2005; Paux et al. 2006), inter-retrotransposon amplified polymorphism (IRAP; Kalendar and Schulman 2006), and retrotransposon-based insertion polymorphism (RBIP; Flavell et al. 1998). In view of this, the whole genome sequence data of foxtail millet were comprehensively analyzed to identify two types of TEs, namely, class I retrotransposons (6314) and class II DNA transposons (24,392) (Yadav et al. 2015a). From 30,706 different TEs, 20,278 repeat junction-based makers were developed, which fall into six types, namely RBIP (4801; ~24%), IRAP (3239; ~16%), RJM (4451; ~22%), RJJM (329; ~2%), ISBP (7401; ~36%), and RMAP (57; 0.2%). Using these TE-based markers, the population structure of 99 foxtail millet accessions was analyzed, which classified these accessions into four groups (A–D). The accessions were further categorized as pure and mixed (admixture) ancestry. The Bayesian model-based population structure corroborated well with NJ tree (Yadav et al. 2015a).

4.3.3 Genes and Gene Families Identified and Analyzed for Trait Improvement

In foxtail millet, not much information is available on the identification of QTLs governing agronomic traits, although Gupta et al. (2014) used 50 SSR markers to analyze the population structure of 184 diverse foxtail millet accessions collected from different eco-geographical regions representing 10 countries. Correlation among 20 agronomic traits evaluated in 184 accessions ($p < 0.05$) revealed wider phenotypic trait variations, which highlighted the suitability of these accessions to constitute an association panel for mapping different traits (Gupta et al. 2014). Association mapping of 20 yield-contributing agronomic traits identified 8 markers associated with 9 different agronomic traits ($p < 0.05$), which contributes to 6–25% of the total phenotypic variation (Gupta et al. 2014). Previously, Doust et al. (2004) identified the four QTLs, each controlling basal branching (tillering) and axillary branching in foxtail millet. Recently, Fang et al. (2016) have constructed a high-density genetic map using 1013 SSRs and identified 29 QTL for 11 agronomic traits. With the advent of NGS, Zhang et al. (2017) have resequenced 439 recombinant inbred lines to develop high-resolution bin map and high-density SNP markers, and have mapped 59 QTLs for 14 agronomic traits.

From a functional genomics perspective, several important genes and their corresponding families were analyzed using computational approaches for elucidating their roles in growth, development and stress response. However, the quantum of research has been performed on clarifying the role of genes in molecular stress response. Among the different stress-responsive transcription factors (TFs), [NAM (no apical meristem), ATAF1 and −2, and CUC2 (cup-shaped cotyledon)] (NAC), and dehydration-responsive element-binding (DREB) are well-studied in foxtail millet for their involvement in salinity and dehydration stress, respectively (Lata et al. 2010; Puranik et al. (2011a, b). In addition, Lata et al. (2013) have identified an SNP linked to dehydration tolerance in the *DREB2* locus of foxtail millet, which was used to develop allele-specific markers useful for breeding purposes. Other than TFs, stress-responsive genes, namely WD40 and 14-3-3, were characterized for their putative roles in stress response (Mishra et al. 2012, 2014; Kumar et al. 2015). Availability of genome sequence and annotation data has facilitated identification of gene families on a genome-wide scale and, in this context, NAC was the first gene family studied (Puranik et al. 2013). A total of 147 genes encoding for NAC TFs were identified in foxtail millet, and 50 candidates were chosen for their expression profiling in response to stress and hormone treatments. The study identified *SiNAC128* as the potential candidate for further functional characterization and stress-associated studies (Puranik et al. 2013). Similarly, AP2/ERF, MYB, C_2H_2 zinc finger, and WRKY TFs were analyzed on a genome-wide scale, and 171, 209, 124, and 110 genes were identified, respectively (Lata et al. 2014; Muthamilarasan et al. 2014b, c, 2015a).

The RNA silencing components, namely Dicer-like (DCL), Argonaute, (AGO) and RNA-dependent RNA polymerase (RDR) were analyzed in foxtail millet genome to identify 8, 19, and 11 genes, respectively (Yadav et al. 2015b). Among these, *SiDCL06*, *SiAGO08*, and *SiRDR07* were pinpointed as potential candidates for further functional studies (Yadav et al. 2015b). In another study, 13 gene families involved in secondary cell wall biosynthesis, namely cellulose synthase, cellulose synthase-like, glucan synthase-like, phenylalanine ammonia lyase, *trans*-cinnamate 4-hydroxylase, 4-coumarate CoA ligase, hydroxycinnamoyl CoA:shikimate/quinate hydroxycinnamoyl transferase, *p*-coumaroyl shikimate 3′-hydroxylase, caffeoyl CoA 3-*O*-methyltransferase, ferulate 5-hydroxylase, caffeic acid *O*-methyltransferase, cinnamoyl CoA reductase, and cinnamyl alcohol dehydrogenase were analyzed and 14, 39, 12, 10, 3, 20, 2, 2, 6, 2, 4, 33, and 13 genes were identified, respectively (Muthamilarasan et al. 2015b; Muthamilarasan and Prasad 2017). Altogether, these studies have highlighted the potential candidate genes which could play a major role in molecular and physiological response to different stresses.

ADP-ribosylation factors (ARFs) have been reported to function in diverse physiological and molecular activities. Recent evidence also demonstrates the involvement of ARFs in conferring tolerance to biotic and abiotic stresses in plant species. In view of this, Muthamilarasan et al. (2016b) studied the ARF proteins in C_3 model rice and C_4 model foxtail millet. A total of 23 and 25 ARF proteins were identified in rice and foxtail millet, respectively. These proteins are classified into four classes (I–IV) based on phylogenetic analysis, with ARFs in classes I–III and ARF-like proteins (ARLs) in class IV. Sequence alignment and domain analysis revealed the presence of conserved and additional motifs, which may contribute to neofunctionalization and subfunctionalization of these proteins. Promoter analysis showed the presence of several *cis*-regulatory elements related to stress and hormone response, indicating their role in stress regulatory network. Expression analysis of rice *ARFs* and *ARLs* in different tissues, stresses, and abscisic acid treatments highlighted temporal and spatial diversification of gene expression. Five rice cultivars screened for allelic variations in *OsARF* genes showed the presence of allelic polymorphisms in a few gene loci. Altogether, the study provided insight into the characteristics of *ARF/ARL* genes in rice and foxtail millet, which could be deployed for further functional analysis to extrapolate their precise roles in abiotic stress responses.

4.4 Online Resources for Foxtail Millet Genomics

With the advances made in genetics and genomics of foxtail millet, several web-based databases have also been constructed to cater for the developed resources to the global research community (Table 4.2). Phytozome is the first online portal (https://phytozome.jgi.doe.gov/pz/portal.html) to host the foxtail millet genome sequenced by JGI.

Table 4.2 List of webtools available for foxtail millet genomics research

Database	URL	Purpose	Reference
Phytozome v12	https://phytozome.jgi.doe.gov/pz/portal.html#!info?alias=Org_Sitalica	Database for browsing the complete information about foxtail millet genome (Bennetzen et al. 2012)	Goodstein et al. (2012)
Gramene v3.0.1	http://ensembl.gramene.org/Setaria_italica/Info/Index		Tello-Ruiz et al. (2016)
PlantGDB	http://www.plantgdb.org/SiGDB/		Duvick et al. (2008)
Foxtail millet database	http://foxtailmillet.genomics.org.cn	Database for browsing the complete information about foxtail millet genome (Zhang et al. 2012)	Zhang et al. (2012)
Setaria italica functional genomics database	http://structuralbiology.cau.edu.cn/SIFGD/	Integrated database of genome, transcript and protein sequences, and miRNA-seq and RNA-seq data from public data sources	You et al. (2015)
Foxtail millet marker database	http://www.nipgr.res.in/foxtail.html	Database of molecular markers developed and the database integrates physical and comparative maps	Suresh et al. (2013)
Foxtail millet miRNA database	http://59.163.192.91/FmMiRNADb/index.html	Database of complete miRNA data identified in foxtail millet along with physical map, comparative maps, target analysis and miRNA-based marker information	Khan et al. (2014)
Foxtail millet transcription factor database	http://59.163.192.91/FmTFDb/index.html	Genome-wide transcription factors identified in foxtail millet along with physical map, tissue-specific expression data and their functional annotation	Bonthala et al. (2014)
Foxtail millet transposable elements-based marker database	http://59.163.192.83/ltrdb/index.html	Database of transposable elements (TEs) identified in foxtail millet along with TE-based markers, physical map and other related information	Yadav et al. (2015a, b)

The v2.2 of the genome consists of the main genome assembly (~405.7 Mb) arranged in 336 scaffolds, and ~400.9 Mb are arranged in 6791 contigs (~1.2% gap). Furthermore, 98.9% of the sequence data is represented in the 9 pseudomolecules, and there were 34,584 loci containing protein-coding transcripts and 43,001 protein-coding transcripts (https://phytozome.jgi.doe.gov/pz/portal.html#!info?alias=Org_Sitalica). Similarly, Gramene (http://www.gramene.org/) and PlantGDB (http://www.plantgdb.org/SiGDB/) host the foxtail millet genome. Several features available within these databases including genome browser, sequence assemblies, special datasets, and other tools facilitate an unrestricted access to the genome sequence information of foxtail millet. The sequence data of Zhang et al. (2012) is available in Foxtail millet Database (http://foxtailmillet.genomics.org.cn); however, the utility of this database is limited because of the availability of updated annotation made available by JGI. Furthermore, application-specific databases, namely Foxtail millet Marker Database (http://www.nipgr.res.in/foxtail.html), Foxtail millet TF Database (http://59.163.192.91/FmTFDb/index.html), Foxtail millet MiRNA Database (http://59.163.192.91/FmMiRNADb/index.html), and Foxtail millet TE-based Marker Database (http://59.163.192.83/ltrdb/index.html), were developed at a later stage to expedite crop improvement of underutilized millets, cereals, and biofuel grasses. These resources are widely used by the researchers and breeders of national and international institutes for structural and functional genomic studies in millets. Being novel resources, the data of markers, genes and miRNAs developed in foxtail millet can certainly assist in the development of resilient crops for climate-smart agriculture.

4.5 Conclusions

India tops the list of global millet production (11.165 million tonnes in 2014); however, pearl millet and finger millet are the top two varieties extensively cultivated in India (Bandyopadhyay et al. 2017). Despite the prominent attributes encompassed by foxtail millet, breeding technologies used in this crop are far less developed than pearl millet and finger millet, and other major cereals such as rice and wheat. In addition, foxtail millet has received a very little research attention worldwide, and less effort has been invested in dissecting the genetic determinants of the prominent traits, which are important for improvement of this model species as well as other millets, cereals, and biofuel crops. In this context, the role of structural genomics is inevitable as it focuses on the physical structure of the genome, aiming to identify, locate, and order the important traits encoded by the chromosomes. DNA-based molecular markers play a vital role in structural genomics as they are imperative for various applications, such as the investigation of genetic diversity and phylogenetic relationships, construction of high-density genome maps, mapping of genes, comparative genome mapping, and marker-assisted selection for crop improvement. Development of genomic resources, including large-scale, genome-wide molecular markers, is the first step toward improving the breeding strategies, which subsequently lead to the development of elite cultivars with desirable agronomic traits. Among the different types of markers, genic markers, namely EST-derived SSRs—EST-SSRs or eSSRs, intron-length polymorphisms (ILPs), and microRNA-based markers, are very useful in understanding marker-trait associations and genomics-assisted breeding for crop improvement. In this context, several marker resources have been developed in foxtail millet and were made available to the global research community through web-based databases. The applicability of these markers in several genotyping purposes including germplasm characterization, transferability, phylogenetics and comparative mapping studies was also established. Altogether, development of molecular markers on a large-scale and demonstration of their utility in mapping important agronomic traits, along with the development of other genomic resources, would be insightful in effecting a new era of foxtail millet breeding.

Acknowledgements Studies on millet genomics in Dr. Manoj Prasad's laboratory are supported by Science and Engineering Research Board (SERB), Department of Science and Technology (DST), Govt. of India [Grant No. EMR/2015/000464], by Department of Biotechnology, Govt. of India [Grant No. BT/HRD/NBA/37/01/2014], and by Core Grant of National Institute of Plant Genome Research (NIPGR), New Delhi, India.

References

Abdelkrim J, Robertson B, Stanton JA, Gemmell N (2009) Fast, cost-effective development of species specific microsatellite markers by genomic sequencing. Biotechniques 46:185–192

Aggarwal RK, Hendre PS, Varshney RK, Bhat PR, Krishna KV, Singh L (2007) Identification, characterization and utilization of EST-derived genic microsatellite markers for genome analyses of coffee and related species. Theor Appl Genet 114:359–372

Andersen JR, Lubberstedt T (2003) Functional markers in plants. Trends Plant Sci 8:554–560

Badoni S, Das S, Sayal YK, Gopalakrishnan S, Singh AK, Rao AR, Agarwal P, Parida SK, Tyagi AK (2016) Genome-wide generation and use of informative intron-spanning and intron-length polymorphism markers for high-throughput genetic analysis in rice. Sci Rep 6:23765

Bandyopadhyay T, Muthamilarasan M, Prasad M (2017) Millets for Next Generation Climate-Smart Agriculture. Frontiers in Plant Science 8

Bennetzen JL (2000) Transposable element contributions to plant gene and genome evolution. Plant Mol Biol 42:251–269

Bennetzen JL, Schmutz J, Wang H, Percifield R, Hawkins J, Pontaroli AC, Estep M, Feng L, Vaughn JN, Grimwood J, Jenkins J, Barry K, Lindquist E, Hellsten U, Deshpande S, Wang X, Wu X, Mitros T, Triplett J, Yang X, Ye CY, Mauro-Herrera M, Wang L, Li P, Sharma M, Sharma R, Ronald PC, Panaud O, Kellogg EA, Brutnell TP, Doust AN, Tuskan GA, Rokhsar D, Devos KM (2012) Reference genome sequence of the model plant *Setaria*. Nature Biotechnol 30:555–561

Boguski MS, Lowe TM, Tolstoshev CM (1993) dbEST-database for "expressed sequence tags". Nat Genet 4:332–333

Bonthala VS, Muthamilarasan M, Roy R, Prasad M (2014) FmTFDb: a foxtail millet transcription factors database for expediting functional genomics in millets. Mol Biol Rep 41:6343–6348

Brachi B, Morris GP, Borevitz JO (2011) Genome-wide association studies in plants: the missing heritability is in the field. Genome Biol 12:232

Castoe TA, Poole AW, Gu W, de Koning APJ, Daza JM, Smith EN, Pollock DD (2010) Rapid identification of thousands of copperhead snake (*Agkistrodon contortrix*) microsatellite loci from modest amounts of 454 shotgun genome sequence. Mol Ecol Resour 10:341–347

Choi HK, Kim D, Uhm T, Limpens E, Lim H, Mun JH, Kalo P, Penmetsa RV, Seres A, Kulikova O, Roe BA, Bisseling T, Kiss GB, Cook DR (2004) A sequence-based genetic map of *Medicago truncatula* and comparison of marker colinearity with *M. sativa*. Genetics 166:1463–1502

Choudhary S, Gaur R, Gupta S, Bhatia S (2012) EST derived genic molecular markers: development and utilization for generating an advanced transcript map of chickpea. Theor Appl Genet 124:1449–1462

Csencsics D, Brodbeck S, Holderegger R (2010) Cost-effective, species-specific microsatellite development for the endangered Dwarf Bulrush (*Typha minima*) using next-generation sequencing technology. J Hered 101:789–793

Devos KM, Wang ZM, Beales J, Sasaki T, Gale MD (1998) Comparative genetic maps of foxtail millet (*Setaria italica*) and rice (*Oryza sativa*). Theor Appl Genet 96:63–68

Devos KM, Ma J, Pontaroli AC, Pratt LH, Bennetzen JL (2005) Analysis and mapping of randomly chosen bacterial artificial chromosome clones from hexaploid bread wheat. Proc Natl Acad Sci U S A 102:19243–19248

Diao X, Schnable J, Bennetzen J, Li J (2014) Initiation of Setaria as a model plant. Front Agr Sci Eng 1:16–20

Doust AN, Devos KM, Gadberry MD, Gale MD, Kellogg EA (2004) Genetic control of branching in foxtail millet. Proc Natl Acad Sci U S A 101:9045–9050

Duvick J, Fu A, Muppirala U, Sabharwal M, Wilkerson MD, Lawrence CJ, Lushbough C, Brendel V (2008) PlantGDB: a resource for comparative plant genomics. Nucl Acids Res 36:D959–D965

Fang X, Dong K, Wang X, Liu T, He J, Ren R, Zhang L, Liu R, Liu X, Li M, Huang M, Zhang Z, Yang T (2016) A high density genetic map and QTL for agronomic and yield traits in *Foxtail millet* [*Setaria italica* (L.) P. Beauv]. BMC Genom 17:336

Flavell AJ, Knox MR, Pearce SR, Ellis TH (1998) Retrotransposon-based insertion polymorphisms (RBIP) for high throughput marker analysis. Plant J 16:643–650

Fu SW, Chen L, Man YG (2011) miRNA biomarkers in breast cancer detection and management. J Cancer 2:116–122

Fu D, Ma B, Mason AS, Xiao M, Wei L, An Z (2013) MicroRNA-based molecular markers: a novel PCR-based genotyping technique in *Brassica* species. Plant Breed 132:375–381

Ganie SA, Mondal TK (2015) Genome-wide development of novel miRNA-based microsatellite markers of rice (*Oryza sativa*) for genotyping applications. Mol Breed 35:51

Goodstein DM, Shu S, Howson R, Neupane R, Hayes RD, Fazo J, Mitros T, Dirks W, Hellsten U, Putnam N, Rokhsar DS (2012) Phytozome: a

comparative platform for green plant genomics. Nucleic Acids Res 40:D1178–D1186
Grady WM, Tewari M (2010) The next thing in prognostic molecular markers: microRNA signatures of cancer. Gut 59:706–708
Grzebelus D (2006) Transposon insertion polymorphism as a new source of molecular markers. J Fruit Ornam Plant Res 14:2006
Gupta PK, Varshney RK (2000) The development and use of microsatellites markers for genetic analysis and plant breeding with emphasis on bread wheat. Euphytica 113:163–185
Gupta S, Kumari K, Das J, Lata C, Puranik S, Prasad M (2011) Development and utilization of novel intron length polymorphic markers in foxtail millet [*Setaria italica* (L.) P. Beauv.]. Genome 54:586–602
Gupta S, Kumari K, Sahu PP, Vidapu S, Prasad M (2012a) Sequence based novel genomic microsatellite markers for robust genotyping purposes in foxtail millet [*Setaria italica* (L.) P. Beauv.]. Plant Cell Rep 31:323–337
Gupta SK, Bansal R, Gopalakrishna T (2012b) Development of intron length polymorphism markers in cowpea [*Vigna unguiculata* (L.) Walp.] and their transferability to other *Vigna* species. Mol Breed 30:1363–1370
Gupta S, Kumari K, Muthamilarasan M, Subramanian A, Prasad M (2013) Development and utilization of novel SSRs in foxtail millet [*Setaria italica* (L.) P. Beauv.]. Plant Breed 132:367–374
Gupta S, Kumari K, Muthamilarasan M, Parida SK, Prasad M (2014) Population structure and association mapping of yield contributing agronomic traits in foxtail millet. Plant Cell Rep 33:881–893
Hamblin MT, Buckler ES, Jannink JL (2011) Population genetics of genomics-based crop improvement methods. Trends Genet 27:98–106
He C, Liu H, Su S, Lu Y, Luo B, Nie Z, Wu L, Liu D, Zhang X, Rong T, Gao S (2015) Genome-wide identification of candidate phosphate starvation responsive genes and the development of intron length polymorphism markers in maize. Plant Breed 134:11–16
Iwata H, Minamikawa MF, Kajiya-Kanegae H, Ishimori M, Hayashi T (2016) Genomics-assisted breeding in fruit trees. Breed Sci 66:100–115
Jaikishan I, Rajendrakumar P, Madhusudhana R, Elangovan M, Patil JV (2015) Development and utility of PCR-based intron polymorphism markers in sorghum [*Sorghum bicolor* (L.) Moench]. J Crop Sci Biotech 18:309–318
Jia XP, Shi YS, Song YC, Wang YG, Wang TY, Li Y (2007) Development of EST-SSR in foxtail millet (*Setaria italica*). Genet Res Crop Evol 54:233–236
Jia X, Zhang Z, Liu Y, Zhang C, Shi Y, Song Y, Wang T, Li Y (2009) Development and genetic mapping of SSR markers in foxtail millet [*Setaria italica* (L.) P. Beauv.]. Theor Appl Genet 118:821–829
Kalendar R, Schulman AH (2006) IRAP and REMAP for retrotransposon-based genotyping and fingerprinting. Nat Protoc 1:2478–2484
Kalendar R, Flavell AJ, Ellis TH, Sjakste T, Moisy C, Schulman AH (2011) Analysis of plant diversity with retrotransposon-based molecular markers. Heredity 106:520–530
Kalia RK, Rai MK, Kalia S, Singh R, Dhawan AK (2011) Microsatellites markers: an overview of the recent progress in plants. Euphytica 177:309–334
Khan Y, Yadav A, Suresh BV, Muthamilarasan M, Yadav CB, Prasad M (2014) Comprehensive genome-wide identification and expression profiling of foxtail millet [*Setaria italica* (L.)] miRNAs in response to abiotic stress and development of miRNA database. Plant Cell Tissue Organ Cult 118:279–292
Kole C, Muthamilarasan M, Henry R, Edwards D, Sharma R, Abberton M, Batley J, Bentley A, Blakeney M, Bryant J, Cai H, Cakir M, Cseke LJ, Cockram J, de Oliveira AC, De Pace C, Dempewolf H, Ellison S, Gepts P, Greenland A, Hall A, Hori K, Hughes S, Humphreys MW, Iorizzo M, Ismail AM, Marshall A, Mayes S, Nguyen HT, Ogbonnaya FC, Ortiz R, Paterson AH, Simon PW, Tohme J, Tuberosa R, Valliyodan B, Varshney RK, Wullschleger SD, Yano M, Prasad M (2015) Application of genomics-assisted breeding for generation of climate resilient crops: progress and prospects. Front Plant Sci 6:563
Kumar K, Muthamilarasan M, Bonthala VS, Roy R, Prasad M (2015) Unraveling 14-3-3 proteins in C_4 panicoids with emphasis on model plant *Setaria italica* reveals phosphorylation-dependent subcellular localization of RS splicing factor. PLoS ONE 10: e0123236
Kumari K, Muthamilarasan M, Misra G, Gupta S, Subramanian A, Parida SK, Chattopadhyay D, Prasad M (2013) Development of eSSR-markers in *Setaria italica* and their applicability in studying genetic diversity, cross-transferability and comparative mapping in millet and non-millet species. PLoS ONE 8:e67742
Lata C, Sahu PP, Prasad M (2010) Comparative transcriptome analysis of differentially expressed genes in foxtail millet (*Setaria italica* L.) during dehydration stress. Biochem Biophy Res Commun 393:720–727
Lata C, Gupta S, Prasad M (2013) Foxtail millet: a model crop for genetic and genomic studies in bioenergy grasses. Crit Rev Biotechnol 33:328–343
Lata C, Mishra AK, Muthamilarasan M, Bonthala VS, Khan Y, Prasad M (2014) Genome-wide investigation and expression profiling of AP2/ERF Transcription factor superfamily in foxtail millet (*Setaria italica* L.). PLoS ONE 9:e113092
Lorenz AJ, Chao S, Asoro FG, Heffner EL, Hayashi T, Iwata H, Smith KP, Sorrells ME, Jannink JL (2011) Genomic selection in plant breeding: knowledge and prospects. Adv Agron 110:77–123
Luce AC, Sharma A, Mollere OS, Wolfgruber TK, Nagaki K, Jiang J, Presting GG, Dawe RK (2006) Precise centromere mapping using a combination of repeat junction markers and chromatin immunoprecipitation-polymerase chain reaction. Genetics 174:1057–1061

Migicovsky Z, Myles S (2017) Exploiting wild relatives for genomics-assisted breeding of perennial crops. Front Plant Sci 8:460

Mishra AK, Puranik S, Bahadur RP, Prasad M (2012) The DNA-binding activity of an AP2 protein is involved in transcriptional regulation of a stress-responsive gene, *SiWD40*, in foxtail millet. Genomics 100:252–263

Mishra AK, Muthamilarasan M, Khan Y, Parida SK, Prasad M (2014) Genome-wide investigation and expression analyses of WD40 protein family in the model plant foxtail millet (*Setaria italica* L.). PLoS ONE 9:e86852

Mondal TK, Ganie SA (2014) Identification and characterization of salt responsive miRNA-SSR markers in rice (Oryza sativa). Gene 535(2):204–209

Muthamilarasan M, Prasad M (2015) Advances in *Setaria* genomics for genetic improvement of cereals and bioenergy grasses. Theor Appl Genet 128:1–14

Muthamilarasan M, Prasad M (2013) Cutting-edge research on plant miRNAs. Curr Sci 104:287–289

Muthamilarasan M, Prasad M (2017) Genetic determinants of drought stress tolerance in *Setaria*. In: Doust A, Diao X (eds) Genetics and genomics of Setaria. Springer, Berlin, pp 267–289

Muthamilarasan M, Suresh BV, Pandey G, Kumari K, Parida SK, Prasad M (2014a) Development of 5123 intron-length polymorphic markers for large-scale genotyping applications in foxtail millet. DNA Res 21:41–52

Muthamilarasan M, Khandelwal R, Yadav CB, Bonthala VS, Khan Y, Prasad M (2014b) Identification and molecular characterization of MYB transcription factor superfamily in C_4 model plant foxtail millet (*Setaria italica* L.). PLoS ONE 9:e109920

Muthamilarasan M, Bonthala VS, Mishra AK, Khandelwal R, Khan Y, Roy R, Prasad M (2014c) C_2H_2-type of zinc finger transcription factors in foxtail millet define response to abiotic stresses. Funct Integr Genom 14:531–554

Muthamilarasan M, Bonthala VS, Khandelwal R, Jaishakar J, Shweta S, Nawaz K, Prasad M (2015a) Global analysis of WRKY transcription factor superfamily in *Setaria* identifies potential candidates involved in abiotic stress signaling. Front Plant Sci 6:910

Muthamilarasan M, Khan Y, Jaishankar J, Shweta S, Lata C, Prasad M (2015b) Integrative analysis and expression profiling of secondary cell wall genes in C_4 biofuel model *Setaria italica* reveals targets for lignocellulose bioengineering. Front Plant Sci 6:965

Muthamilarasan M, Dhaka A, Yadav R, Prasad M (2016a) Exploration of millet models for developing nutrient rich graminaceous crops. Plant Sci 242:89–97

Muthamilarasan M, Mangu VR, Zandkarimi H, Prasad M, Baisakh N (2016b) Structure, organization and evolution of ADP-ribosylation factors in rice and foxtail millet, and their expression in rice. Sci Rep 6:24008

Palumbi SR (1995) Nucleic acids II: the polymerase chain reaction. In: Hillis D, Moritz C (eds) Molecular systematics, 2nd edn. Sinauer Associates Inc., Sunderland, pp 205–247

Pandey G, Misra G, Kumari K, Gupta S, Parida SK, Chattopadhyay D, Prasad M (2013) Genome-wide development and use of microsatellite markers for large-scale genotyping applications in foxtail millet [*Setaria italica* (L.)]. DNA Res 20:197–207

Panjabi P, Jagannath A, Bisht NC, Padmaja KL, Sharma S, Gupta V, Pradhan AK, Pental D (2008) Comparative mapping of *Brassica juncea* and *Arabidopsis thaliana* using Intron Polymorphism (IP) markers: homoeologous relationships, diversification and evolution of the A, B and C Brassica genomes. BMC Genom 9:113

Paux E, Roger D, Badaeva E, Gay G, Bernard M, Sourdille P, Feuillet C (2006) Characterizing the composition and evolution of homoeologous genomes in hexaploid wheat through BAC-end sequencing on chromosome 3B. Plant J 48:463–474

Pazhamala L, Saxena RK, Singh VK, Sameerkumar CV, Kumar V, Sinha P, Patel K, Obala J, Kaoneka SR, Tongoona P, Shimelis HA, Gangarao NVPR, Odeny D, Rathore A, Dharmaraj PS, Yamini KN, Varshney RK (2015) Genomics-assisted breeding for boosting crop improvement in pigeonpea (Cajanus cajan). Front Plant Sci 6:50

Puranik S, Bahadur RP, Srivastava PS, Prasad M (2011a) Molecular cloning and characterization of a membrane associated NAC family gene, SiNAC from foxtail millet [*Setaria italica* (L.) P. Beauv.]. Mol Biotechnol 49:138–150

Puranik S, Jha S, Srivastava PS, Sreenivasulu N, Prasad M (2011b) Comparative transcriptome analysis of contrasting foxtail millet cultivars in response to short-term salinity stress. J Plant Physiol 168:280–287

Puranik S, Sahu PP, Mandal SN, Suresh BV, Parida SK, Prasad M, Zhang T (2013) Comprehensive genome-wide survey, genomic constitution and expression profiling of the NAC transcription factor family in foxtail millet (Setaria italica L.). PLoS ONE 8(5):e64594

Saha D, Gowda MVC, Arya L, Verma M, Bansal KC (2016) Genetic and genomic resources of small millets. Crit Rev Plant Sci 35:56–79

Schmitt BA, Costa JH, de Melo DF (2006) AOX-A functional marker for efficient cell reprogramming under stress? Trends Plant Sci 11:281–287

Shu Y, Li Y, Zhu Y, Zhu Z, Lv D, Bai X, Cai H, Ji W, Guo D (2010) Genome-wide identification of intron fragment insertion mutations and their potential use as SCAR molecular markers in the soybean. Theor Appl Genet 121:1–8

Suresh BV, Muthamilarasan M, Misra G, Prasad M (2013) FmMDb: a versatile database of foxtail millet markers for millets and bioenergy grasses research. PLoS ONE 8:e71418

Tello-Ruiz MK, Stein J, Wei S, Preece J, Olson A, Naithani S, Amarasinghe V, Dharmawardhana P, Jiao Y, Mulvaney J, Kumari S, Chougule K, Elser J, Wang B, Thomason J, Bolser DM, Kerhornou A, Walts B, Fonseca NA, Huerta L, Keays M, Tang YA, Parkinson H, Fabregat A, McKay S, Weiser J,

D'Eustachio P, Stein L, Petryszak R, Kersey PJ, Jaiswal P, Ware D (2016) Gramene 2016: comparative plant genomics and pathway resources. Nucleic Acids Res 44:D1133–D1140

Upadhyaya HD, Pundir RPS, Gowda CLL, Reddy VG, Singh S (2008) Establishing a core collection of foxtail millet to enhance the utilization of germplasm of an underutilized crop. Plant Genet Resour Charact Util 7:177–184

Varshney RK, Thudi M, Aggarwal R, Börner A (2007) Genic molecular markers in plants: development and applications. In: Varshney RK, Tuberosa R (eds) Genomics-assisted crop improvement: genomics approaches and platforms (vol 1). Springer, Dordrecht, pp 13–29

Wang ZM, Devos KM, Liu CJ, Wang RQ, Gale MD (1998) Construction of RFLP-based maps of foxtail millet, *Setaria italica* (L.) P. Beauv Theor Appl Genet 96:31–36

Wang X, Zhao X, Zhu J, Wu W (2005) Genome-wide investigation of intron length polymorphisms and their potential as molecular markers in rice (*Oryza sativa* L.). DNA Res 12:417–427

Wang CB, Guo WZ, Cai CP (2006) Characterization, development and exploitation of EST-derived microsatellites in *Gossypium raimondii* Ulbrich. Chin Sci Bull 51:316–320

Wang Y, Chen J, Francis DM, Shen H, Wu T, Yang W (2010) Discovery of intron polymorphisms in cultivated tomato using both tomato and *Arabidopsis* genomic information. Theor Appl Genet 121:1199–1207

Wanjugi H, Coleman-Derr D, Huo N, Kianian SF, Luo MC, Wu J, Anderson O, Gu YQ (2009) Rapid development of PCR-based genome-specific repetitive DNA junction markers in wheat. Genome 52:576–587

Yadav CB, Muthamilarasan M, Pandey G, Prasad M (2014) Development of novel microRNA-based genetic markers in foxtail millet for genotyping applications in related grass species. Mol Breed 34:2219–2224

Yadav CB, Bonthala VS, Muthamilarasan M, Pandey G, Khan Y, Prasad M (2015a) Genome-wide development of transposable elements-based markers in foxtail millet and construction of an integrated database. DNA Res 22:79–90

Yadav CB, Muthamilarasan M, Pandey G, Prasad M (2015b) Identification, characterization and expression profiling of Dicer-like, Argonaute and RNA-dependent RNA polymerase gene families in foxtail millet. Plant Mol Biol Rep 33:43–55

You FM, Wanjugi H, Huo N, Lazo GR, Luo MC, Anderson OD, Dvorak J, Gu YQ (2010) RJPrimers: unique transposable element insertion junction discovery and PCR primer design for marker development. Nucleic Acids Res 38:W313–W320

You Q, Zhang L, Yi X, Zhang Z, Xu W, Su Z (2015) SIFGD: *Setaria italica* functional genomics database. Mol Plant 8:967–970

Zhang G, Liu X, Quan Z, Cheng S, Xu X, Pan S, Xie M, Zeng P, Yue Z, Wang W, Tao Y, Bian C, Han C, Xia Q, Peng X, Cao R, Yang X, Zhan D, Hu J, Zhang Y, Li H, Li H, Li N, Wang J, Wang C, Wang R, Guo T, Cai Y, Liu C, Xiang H, Shi Q, Huang P, Chen Q, Li Y, Wang J, Zhao Z, Wang J (2012) Genome sequence of foxtail millet (*Setaria italica*) provides insights into grass evolution and biofuel potential. Nature Biotechnol 30:549–554

Zhang K, Fan G, Zhang X, Zhao F, Wei W, Du G, Feng X, Wang X, Wang F, Song G, Zou H, Zhang X, Li S, Ni X, Zhang G, Zhao Z (2017) Identification of QTLs for 14 agronomically important traits in *Setaria italica* based on SNPs generated from high-throughput sequencing. G3 (Bethesda) 7:1587–1594

Zhou R, Jia J, Gao L (2010) RGA-ILP, a new type of functional molecular markers in bread wheat. Euphytica 172:263–273

Zhu H, Senalik D, McCown BH, Zeldin EL, Speers J, Hyman J, Bassil N, Hummer K, Simon PW, Zalapa JE (2012) Mining and validation of pyro-sequenced simple sequence repeats (SSRs) from American cranberry (*Vaccinium macrocarpon* Ait.). Theor Appl Genet 124:87–96

Breeding Strategies in Foxtail Millet

5

K. Hariprasanna, Jinu Jacob, Parashuram Patroti
and K.B.R.S. Visarada

Abstract

Foxtail millet is a highly self-pollinated crop, and the small delicate flowers make hybridization and crossing difficult. Progress in crop improvement of foxtail millet has been achieved chiefly through pedigree selection in many parts of the globe. China has been pioneering in developing male sterile lines and yield gains through heterosis. Many male sterile systems have been developed in China and are used for commercial production. Methods of crossing through physical and chemical treatments are in their infancy. In order to realize a yield benefit in this nutritionally rich and highly climate-resilient crop, recombination breeding and hybrid technology need to be developed.

5.1 Introduction

Foxtail millet is a diploid ($2n = 18$), C4 panicoid crop. Cultivation of foxtail millet is now limited to certain pockets, and in several areas it has been replaced by other crops with irrigation. Its superior nutritional quality coupled with its low requirement for water, makes it a climate-resilient crop suitable for cultivation under dry land agricultural systems (Muthamilarasan and Prasad 2015; Muthamilarasan et al. 2016). It has a small genome, and its use as a model crop for bioenergy has created a momentum with more groups working than before (Muthamilarasan and Prasad 2015). However, unless yield gains are realized it is difficult for farmers to afford to grow this crop. Yield gains through heterosis and recombination have progressed in China much more than other parts of the globe. In other parts of the world the crop improvement in foxtail millet has been achieved to a larger extent by selection and through recombination breeding to a smaller extent. Floral morphology and flowering behavior of this crop make it difficult to take up crosses between the desired parents. Thus, we have seen many research publications to date on developing methods for crossing in foxtail millet. In this chapter we discuss the floral biology, crossing

K. Hariprasanna · J. Jacob · P. Patroti
K.B.R.S.Visarada (✉)
ICAR-Indian Institute of Millets Research,
Rajendranagar, Hyderabad 500030, India
e-mail: visarada@millets.res.in

M. Prasad (ed.), *The Foxtail Millet Genome*, Compendium of Plant Genomes,
DOI 10.1007/978-3-319-65617-5_5

methods, and development of cytoplasmic male sterile (CMS) lines.

5.2 Floral Biology

Floral morphology and anthesis behavior make foxtail millet one of the most difficult species to cross-pollinate (Baltensperger 1996). A floral description of foxtail millet has been presented by Gupta et al. (2011). It is mostly a self-pollinated crop with cross-pollination averaging about 1.4–4% (Li et al. 1935; Till-Bottraud et al. 1992). The inflorescence of foxtail millet is a terminal spike that consists of the main stalk and short side branches (Fig. 5.1a). Branches bear spikes and bristles (Fig. 5.1b, c). Bristles generally terminate inflorescence axes and appear paired with spikelets (Fig. 5.1d, e). There are up to eight orders of branching in foxtail millet, each branch initiating meristems which differentiate into spikelets or bristles (Doust et al. 2005). Each spikelet contains two florets embraced by a pair of glumes (Fig. 5.1f). The lower floret is sterile whereas the upper one is fertile or bisexual with three stamens and a long oval, smooth ovary with two long styles that terminate in a brush-like stigma (Fig. 5.1g) (Hector 1936; Nirmalakumari and Vetriventhan 2010). The anthers are yellow or white in color. Flowering in foxtail millet starts from the top of the main spike and proceeds downward. It takes

Fig. 5.1 Floral biology of foxtail millet

8–16 days for an ear head to complete flowering. A single floret remains open for about 30 min, and it takes around 80 min for complete blooming to take place. The rates of both these parameters are affected by temperature and atmospheric humidity (Heh et al. 1937). During pollination, stigmatic branches emerge first, followed by the emergence of anthers through the slit between palea. Once the anthers are fully extruded outside the glumes, they shed pollen. After dehiscence the glumes close, leaving the shrivelled anthers and tips of stigmas outside. Anthesis in foxtail millet takes place near midnight and between 8 and 10 a.m. A maximum number of florets open on the 6th day of flowering (Sundararaj and Thulasidas 1976). The flowering pattern is influenced by temperature and varies between crop seasons. We found that crops sown in the post-rainy season (*rabi*) flower till 9 a.m. in the cool winter days of December and January. High pollen fertility was observed during this period. This is best time for taking up crosses in India at Hyderabad.

Green foxtail millet (*Setaria viridis*) is a close relative of foxtail millet. It has a short life cycle, small plant stature, and a high number of seed set, and is a model species for the study of C4 plants. Studies have been extensively carried out on understanding its floral biology in this species. Using a time-lapse digital recording technique, Rizal et al. (2013) found that flowering in *Setaria* was triggered by the darkness of the night and a temperature lower than 35 °C. The anthesis of all the spikelets in a panicle took up to three nights flowering from 9:30 p.m. to 10:00 a.m. Flowering started from the tip and proceeded to the middle in the first night. Each spikelet was found to have three phases of anthesis during which pollination occurs. A spikelet generally remained open for less than 3 h. The pollination time for each spikelet was less than 60 min. All spikelets on the cluster did not flower at the same time. The anthers elongated taller than the stigma, and after a rapid desiccation they collapsed, mostly on the stigma. Once the spikelets are closed, vestiges of the anthers and stigma remain outside. The remains of the anthers which turn from yellow to brown are a sign that the spikelets have opened and fertilized. Information from studies on green foxtail millet would enable the geneticists and plant breeders to develop efficient crossing techniques in *Setaria italica*.

5.3 Crop Improvement

Wide genetic diversity is available in the germplasm resources for agronomically important traits, but has remained unexplored (Upadhyaya et al. 2015). Hybridization and selection are primary means of availing genetic variability for cultivar development in any crop species. A very low degree of natural cross-pollination was observed in foxtail millet. The International Crops Research Institute for the Semi-Arid Tropics genebank is presently holding 1,474 cultivated germplasm accessions from 23 countries. To facilitate the breeding efforts, a core collection (10% of the entire collection) was characterized using the taxonomic and qualitative traits. The germplasm accessions were stratified into three taxonomic races (Indica, Maxima, and Moharia) (Upadhyaya et al. 2009). Vetriventhan (2011) identified trait-specific accessions based on their performance in three environments in foxtail millet for economically important traits such as yield and traits contributing to yield (15 accessions for each trait) to be used in recombination breeding to develop high-yielding cultivars. Population diversity was analyzed using SSR (simple sequence repeats) markers and was estimated to be having low linkage disequilibrium (LD), suggesting the possibility of high-resolution association mapping (Wang et al. 2013; Jia et al. 2013; Zhang et al. 2014). This produces ready-to-use information for foxtail millet breeders.

Foxtail millet genotypes under cultivation in India, USA, and many other places, barring China, are mostly selections from landraces, and efforts to avail recombination breeding and hybrid programs are very few. Primarily it is because of the difficulty in making crosses, the florets are very small in size to handle, and flowering takes place in the early hours of the day, making it difficult to work. High heterosis for grain yield and other important agronomic traits was observed in F_1, and the genes controlling grain yield, plant height, and spike length are tightly linked (Siles et al. 2004). Based on the high degree of heterosis observed in an F_2 generation, the authors suggested an alternative of growing F_2 generations or other types of populations with a relatively high percentage of heterozygous genotypes to provide significant yield benefits over non-hybrid varieties. Radiation and chemical-induced mutations have also been used in foxtail millet breeding to create novel types, such as dwarf lines. Single plants were selected from landraces, improved varieties of farmers, and the progeny were tested. Superior progenies, mostly for earliness, pest and disease resistance, and grain yield, were evaluated in multi-location trials, and released as varieties. Genetics of several agronomic characters has been studied and consolidated for ready use in plant breeding programs (Hariprasanna 2017). Pure line selection has resulted in the development and release of a maximum number of foxtail millet varieties in India. Hybridization-based pedigree selection is the main breeding strategy in China. Later, the introduction of male sterility and heterosis contributed to the quantum jump in the yields of foxtail millet in China.

Exploitation of heterosis in crops is the most important means of crop improvement. Heterotic vigor obtained from the cross-fertilization of genetically different parental lines is one of the most important means of improving yield parameters and quality traits, in addition to developing tolerance to stresses in crop plants. Heterosis and hybrid vigor have been demonstrated in China (Diao and Jia 2017). Siles et al. (2004) reported a high level of heterosis for grain yield among intervarietal crosses. The authors suggested that the crosses between highly heterozygous parents provide significant yield gains. China has attained the yield gains through commercial hybrids that were produced based on the male sterile line. Male sterility such as genetic, cytoplasmic, photosensitive or thermosensitive can all be used to make crosses between diverse parents and exploit the recombination or heterosis. Male sterility can be used for the production of commercial hybrids on a large scale and also to recombine diverse parents for genetic gains. All the above types of male sterile lines are generated and used in China, but other parts of the world do not have them. Access to these lines from one country to another is difficult because of intellectual property rights on material transfer. In India, the crop improvement in foxtail millet was limited to pure-line selection, and efforts toward recombination breeding are on increasing trend (Hariprasanna 2017). In the USA the breeding program is limited and most of the lines cultivated today are selections from landraces rather than designed crosses (Siles et al. 2001).

5.4 Crossing Methods

Genetic improvement in this crop suffers because of the absence of an efficient crossing technique. Taking up crosses in foxtail millet is difficult, and hence not exploited to its full potential. Floral morphology and anthesis behavior make foxtail millet [*S. italica* (L.) Beauv.] one of the most difficult species to cross-pollinate (Baltensperger 1996). The minuteness of the flowers, the delicate and environment-dependent process of anthesis, and timing of anthesis make hybridization a tedious task in foxtail millet. Siles et al. (2001) devised a crossing method in which an average 75% seed set and more than 90% true hybrid seed was accomplished by emasculating a high number of florets as optimized on the second day, starting from the opening of flowers. In some of the earliest attempts to enhance the chances of natural cross-pollination, panicles of selected plants were enclosed in a parchment bag before flowering (contact method) (Ayyangar 1934), resulting in a low frequency (1.5%) of true hybrids (Mahishi et al. 1982). The technique of controlled hybridization with the removal of anthers and artificial pollination, although very difficult because of the small flower, started yielding improved cultivars by the mid-1990s.

During emasculation, the bristles of the female and the male parents are excised gently with a pair of scissors. After the bristles are excised, emasculation has to be done when the first anther had just emerged and before the pollen sacs burst. Anthers are to be removed by gently inserting forceps at each side of the palea and pushing the anther out quickly. Each emasculated flower is then immediately marked with a fine point black or blue permanent marker so that emasculated flowers can be identified easily. The male and the female spikelets are tied together or enclosed in a bag. The bag has to be tapped in the early hours of the day for 3 days more. Later the male spikelet is removed, and the female is bagged separately. It is critical to watch and remove the anthers after emergence but before the burst. Jiang et al. (2013) provided a video on the methods of crossing *S. viridis*, which is very useful for setting up a crossing program.

Artificial emasculation is done through a spray or dip in hot water, cold water, and chemical agents. Rizal et al. (2013) found that a floral dip in hot water at 48 °C for 3–6 min was useful to recover three to five outcross progeny per panicle in *S. viridis*. Chemical spray with 500 μM maleic hydrazide was effective without loss of stigma receptivity (Rizal et al. 2015). The authors provided a detailed description of the emasculation procedure, which can be readily followed. In our study with *S. italica* we completely remove the bristles with scissors. Spikelets are removed carefully at random at the base, leaving a few (can be 10–30) florets. Because of this, all the intact spikelets are separated and exposed. Top and bottom spikelets are also removed in the panicle. Panicles for crossing are covered by a small parchment cover, taking care not to break them. We also observed that floral dip gives better results than floral sprays. For crossing, pollen from the male parent was sprayed on the emasculated panicle for crossing (Rizal et al. 2015). We bring the male and female parents together, gently tie the emasculated panicle and pollen shedding panicle with a thread so that they are intact, and later cover them with a paper bag. This ensures pollination if the flowering takes place at odd hours. *S. italica* is the cultivated crop species and has been improved for grain and related traits. Hence it contains long panicles with many secondary branches and more florets (Fig. 5.1b). Thus, emasculation in *S. italica* is more complicated. Flowering time varies with seasons between midnight in summer and the early hours in winter. The crossing programs can be taken up on a large scale during the winter months. Spray of SQ1, a chemical hybridization agent (CHA) at the early protogyny stage at a concentration of 5 kg/ha was effective in the induction of male sterility in the field (Yu-Long et al. 2011; Zhang et al. 2017). This can be used for hybrid seed production on a commercial scale.

5.5 Male Sterility

China has been pioneering the development of male sterile lines of foxtail millet. Utilization of heterosis has gained momentum in China and various male sterile lines have been identified such as (1) genic highly male-sterility lines (GMS) (Hu et al. 1986), (2) photo-(thermo-) sensitive nuclear lines (PMS or TMS) (Cui et al. 1979; Wang et al. 1993, 2002; Zhao et al. 1996; Hao et al. 2009), (3) cytoplasmic male sterility (CMS) (Zhu et al. 1991), (4) cytoplasmic-nuclear male sterile type (Zhi et al. 2007), and (5) partial genetic male sterile line (PAGMS). Research using heterosis for foxtail millet began in the 1960s with the development of male sterile lines by various approaches (Diao and Jia 2017). Production of foxtail millet in China has passed through pedigree selection, development of male sterile lines, and commercialization of hybrids through the use of a partial male sterile line. Many GMS lines were developed through hybridization between landraces and crossing foxtail millet with other *Setaria* species (Cui et al. 1979; Zhu et al. 1991).

A GMS line, Ch78182, was derived from the cross between Australian and Tulufan races and a complete (100%) restorer line, 181–5 and was identified for seed production (Hu et al. 1986). Ch__ genotype is a dominant male sterile gene, and fertility in restorer line is controlled by the epistatic interaction of Ms__ and Rf__ genes. Thus, the male sterility is suppressed completely by the Rf__ gene. In the line Ch78182, sterility was found to be because of the lack of anther dehiscence, though the pollen grain development was complete. Thus, occasional fertility was observed in this line and could not be used for commercial seed production.

Photoperiod-sensitive male sterility (PMS) is a useful genetic tool for the development of two-line hybrids in self-pollinated crops. In this system, the plant fertility/sterility is regulated by the photoperiod, and the PMS gene(s) cause(s) male sterility under long daylight (LD) or short daylight (SD) conditions and results in fertility under an SD or an LD condition. The first PMS line was reported in foxtail millet in 1996, line 821 (Zhao et al. 1996). The PMS line is maintained and multiplied by self-pollination under SD conditions, and hybrid F_1 seeds can be produced through outcrossing with restorer lines under LD conditions. The PMS line had the advantage of obtaining a high purity of F_1 seeds, and line 821 had the disadvantage that it required a critical photoperiod for hybrid production. This lacuna was overcome by the development of a PMS line, JG1S, which was completely sterile under natural LD conditions and fertile under SD conditions. The line was studied in depth and the inheritance pattern analyzed a few years later (Yuan et al. 2008). A PMS line, JG1S, was identified that was completely sterile under long daylight conditions (14.5 h/day) and partially sterile under short daylight conditions (10 h/day) (Yuan et al. 2008).

A highly nuclear-male-sterile line of foxtail millet with the recessive nuclear sterile gene, with 100% of the sterility percentage, and 95% of the sterility degree rate was crossed with a superior restorer line, and six varieties and a series of variant generation materials with favourable characters were bred from the cross (Wang et al. 1993). A highly genic male sterile line Gao146A was identified by Wang et al. (2013), which showed 95% sterile rate and was not sensitive to light and temperature. F_1 generation was fertile, and the segregation ratio of fertility to sterility was 3:1 in F_2 generation, indicating that a single recessive gene controlled the trait. Using F_2 population derived from the cross Gao146A/K103, one gene controlling the highly male sterility, tentatively named *ms1*, was mapped on chromosome VI using SSR markers (Wang et al. 2013). Heterosis has also been used in developing hybrid cultivars. Zhangzagu5, a hybrid cultivar, was released from Zhangjiakou Academy of Agricultural Sciences, Hebei Province, China, and yielded 12,159 kg/ha vs conventional cultivars ranging from 4,500 to 6,000 kg/ha in 2007 (Liu et al. 2014).

CMS is maternally inherited and can be transferred to different genetic backgrounds through repeated backcrossing. Sterility of the lines is maintained by crossing the MS line (A line) to the complimentary fertile (B) line. Suitable restorer lines that restore 100% fertility and show hybrid vigor are used as male parents for commercial hybrid seed production. These CMS lines are developed by different methods. One of them is through spontaneous mutation and/or interracial crosses between geographically distant species. In foxtail millet CMS line derived thus could not lead to true CMS, which was a discouraging factor for commercial seed production. By distant hybridization of *Setaria verticillata* with foxtail millet, a CMS line was developed (Zhu et al. 1991). However, it was not used for hybrid seed production. A significant step in heterosis breeding in foxtail millet was the development of Suanxi 28, a partial male sterile line developed from a spontaneous mutant from the landrace Suanpibai, which led to a new hybrid seed production system (Diao 2017). This line was used for hybrid seed production and was maintained through selfing (3–5% seed set). Thus, a two-line system of hybrids was developed in China with the help of Suanxi 28. Despite the development of several male sterile lines, PAGMS is used successfully in hybrid production (Diao and Jia 2017). New hybrid cultivars developed through the two-line system using PAGMS showed higher yields than conventional varieties. This system prevailed in China for hybrid seed production for many years, even though other types of male sterile lines have been explored and developed.

Interspecific crosses between foxtail millet and green foxtail have resulted in 65–70% sterility, which has been utilized for developing male sterile lines. However, most of the Chinese spring foxtail millet male sterile lines were derived from 'Chang 10A', whose cytoplasm was contributed by 'Qinyuanmujizui' (Liu et al. 2014; Wang et al. 1998). The summer foxtail millet male sterile lines were derived from 'Huangmi 1A' with the cytoplasm from 'Dahuanggu' (Liu et al. 1996, 2006). CMS is inherited maternally, and hence all the hybrid plants carry the same cytoplasm; it threatens the vulnerability of all the female parental lines to CMS-related diseases and disasters. The narrow genetic base of CMS in China was confirmed through analysis using mtDNA-specific primers (Liu et al. 2014). Because of the maternal inheritance of cytoplasm, using a single source of cytoplasm in male sterile lines makes hybrid cultivars vulnerable to infection by cytoplasm-related diseases, leading to epidemics. Thus, there is always interest in searching for novel sources of male sterile lines.

5.6 Interspecific Crosses

The earliest efforts in wide hybridization were successful with *S. viridis* and *Setaria faberi* (Li et al. 1944). The wild green foxtail, *S. viridis*, and the cultivated foxtail millet, *S. italica*, are compatible for crossing at very low rates (Till-Bottraud et al. 1992). A wild relative of cultivated foxtail millet, *S. viridis*, is a source of important traits. These two are compatible, and the genetic background of cultivated line can be easily restored in two backcrosses and the weedy traits in the inter-specific cross derivatives can be eliminated (Naciri et al. 1992; Zangre and Darmency 1993). Darmency et al. (1987) could obtain a twofold increase in seed weight in the F_1 interspecific hybrid after colchicine-induced tetraploidization of F_1. Although the option is promising, it has practical difficulties. From the hybridization of *S. italica* with its wild relative *S. verticillata* (a tetraploid), a CMS line was developed (Zhu et al. 1991). Sterile F_1 plants were backcrossed to *S. italica* in three backcrosses. The Chinese CMS was derived with great difficulty from repeated crosses with *S. verticillata* and hence named as VeCMS. Despite the difficulties in crossing, primary trisomics of foxtail were constituted by inducing autotetraploids in tissue culture and crossing them to the diploid parent (Wang et al. 1999).

5.7 Conclusion

Foxtail millet is nutritionally rich, and entrepreneurship in the nutraceutical industry is picking up globally because of its health benefits. Consistent and constant supply of grain for various end uses in terms of quantity and quality is the prerequisite to meet industrial demands, and it can be met only through increasing yield. Because heterosis, in terms of yield gain, is promising, two areas need to be promoted: (1) recombination breeding and release of high yielding varieties and (2) hybrids for heterosis on a commercial scale. This would contribute to the nutritional security and health security of humans and animals.

References

Ayyangar GNR (1934) Recent work in the genetics of millets in India. Madras Agric J 22:16–26

Baltensperger DD (1996) Foxtail and proso millet. In: Janick J (ed) Progress in new crops. ASHS Press, Alexandria, VA, pp 182–190

Cui WS, Du G, Zhao ZH (1979) The selection and utilization of male sterile line Shanxi 280 of foxtail millet. Sci Agric Sin 12:43–46

Darmency H, Ouin C, Pernes J (1987) Breeding foxtail millet (*Setaria italica*) for quantitative traits after interspecific hybridization and polyploidization. Genome 29:453–456

Diao X (2017) Production and genetic improvement of minor cereals in China. Crop J 5:103–114

Diao X, Jia G (2017) Foxtail millet breeding in China. In: Doust A, Diao X (eds) Genetics and genomics of *Setaria*. Springer International Publishing, Cham, pp 93–113

Doust AN, Devos KM, Gadberry MD, Gale MD, Kellogg EA (2005) The genetic basis for inflorescence variation between foxtail and green millet (Poaceae). Genetics 169:1659–1672

Gupta A, Sood S, Agrawal OK, Bhatt JC (2011) Floral biology and pollination system in small millets. Eur J Plant Sci Biotechnol 6:81–86

Hao XF, Wang JZ, Wang GQ, Wang LY, Wang XY (2009) AFLP analysis of photo sensitive male sterile gene in millet. J Shanxi Agri Sci 37:10

Hariprasanna K (2017) Foxtail millet, *Setaria italica* (L.) P. Beauv. In: Patil JV (ed) Millets and sorghum: biology and genetic improvement, 1st edn. Wiley, UK, pp 112–149. ISBN 978-1-119-12305-7

Hector JH (1936) Introduction to the botany of field crops. In Millets vol I, Cereals. Central New Agency, Johannesburg, South Africa, pp 307–319

Heh CM, Mei TF, Yang SS (1937) Anthesis of millet, *Setaria italica* (L.) Beauv. J Am Soc Agron 29:845–853

Hu HK, Ma SY, Shi YH (1986) The discovery of a dominant male-sterile gene in millet (*Setaria italica*). Acta Agron Sin 12:73–78

Jia G, Huang X, Zhi H, Zhao Y, Zhao Q, Li W, Chai Y, Yang L, Liu K, Lu H, Zhu C, Lu Y, Zhou C, Fan D, Weng Q, Guo Y, Huang T, Zhang L, Lu T, Feng Q, Hao H, Liu H, Lu P, Zhang N, Li Y, Guo E, Wang S, Wang S, Liu J, Zhang W, Chen G, Zhang B, Li W, Wang Y, Li H, Zhao B, Li J, Diao X, Han B (2013) A haplotype map of genomic variations and genome-wide association studies of agronomic traits in foxtail millet (*Setaria italica*). Nat Genet 45:957–961

Jiang H, Barbier H, Brutnell T (2013) Methods for performing crosses in *Setaria viridis*, a new model system for the grasses. J Vis Exp 80:50527

Li HW, Li CH, Pao WK (1944) Cytological and genetical studies of the interspecific cross of the cultivated foxtail millet, *Setaria italica* (L.) Beauv., and the green foxtail millet, *S. viridis* L. J Am Soc Agron 9:32–54

Li HW, Meng CJ, Liu TN (1935) Problems in the breeding of millet [*Setaria italica* (L.) Beauv.]. J Am Soc Agron 27: 693–670

Liu ZL, Cheng RH, Li XY (1996) The pedigree analysis and evaluation of north China summer millets. Crops 5:24

Liu ZL, Cheng RH, Zhang FL, Xia XY, Shi ZG, Hou SL (2006) Millet variety in boreali-sinica summer millets region and its pedigree evolution and analysis on genetic foundation. Acta Agric Boreali-Sin 21:103–109

Liu Z, Zhang T, Li C, Bai G (2014) Genetic diversity and classification of cytoplasm of Chinese elite foxtail millet [*Setaria italica* (L.) P. Beauv.] Germplasm. Crop Sci 54:659–666

Mahishi DM, Aradhya KM, Seetharam A, Gowda BTS (1982) Efficacy of contact method for hybridization in foxtail millet. MILWAI Newslett 1:16

Muthamilarasan M, Prasad M (2015) Advances in *Setaria* genomics for genetic improvement of cereals and bioenergy grasses. Theor Appl Genet 128:1–14

Muthamilarasan M, Dhaka A, Yadav R, Prasad M (2016) Exploration of millet models for developing nutrient rich graminaceous crops. Plant Sci 242:89–97

Naciri Y, Darmency H, Belliard J, Dessaint F, Pernès J (1992) Breeding strategy in foxtail millet, *Setaria italica* (L. P. Beauv.), following interspecific hybridization. Euphytica 60:97

Nirmalakumari A, Vetriventhan M (2010) Characterisation of foxtail millet germplasm collections for yield contributing traits. Elect J Plant Breed 1:140–147

Rizal G, Acebron K, Mogul R, Karki S, Larazo N, Quick WP (2013) Study of flowering pattern in Setaria viridis, a proposed model species for C4 photosynthesis research. J Bot 2013:7

Rizal G, Karki S, Garcia R, Larazo N, Alcasid M, Quick WP (2015) The use of maleic hydrazide for effective hybridization of *Setaria viridis*. PLoS ONE 10:e0125092

Siles MM, Baltensperger DD, Nelson LA (2001) Technique for artificial hybridization of foxtail millet [*Setaria italica* (L.) Beauv.]. Crop Sci 41:1408–1412

Siles MM, Russell WK, Nelson LA, Baltensperger DD, Johnson B, Van Vleck LD, Jensen SG, Hein GL (2004) Heterosis for grain yield and other agronomic traits in foxtail millet. Faculty Papers and Publications in Animal Science, Paper, p 156

Sundararaj DP, Thulasidas G (1976) Botany of field crops. Mac Millan publishers, India, p 509

Till-Bottraud I, Reboud X, Brabant P, Lefranc M, Rherissi B, Vedel F, Darmency H (1992) Out-crossing and hybridization in wild and cultivated foxtail millets: consequences for the release of transgenic crops. Theor Appl Genet 83:940–946

Upadhyaya HD, Pundir RPS, Gowda CLL, Gopal Reddy V, Singh S (2009) Establishing a core collection of foxtail millet to enhance the utilization of germplasm of an underutilized crop. Plant Genet Res 7:177–184

Upadhyaya H, Vetriventhan M, Deshpande SP, Sivasubramani S, Wallace JG, Buckler ES, Hash CT, Ramu P (2015) Population genetics and structure of a global foxtail millet germplasm collection. Plant Genome 8:1–13

Vetriventhan M (2011) Phenotypic and genetic diversity in the foxtail millet (*setaria italica* (l.) P. Beauv.) core collection). PhD thesis, Tamil Nadu Agricultural University

Wang T, Du RH, Hao F (1993) Studies and utilization of highly male sterility of summer millet (*Setaria italica*). Sci Agric Sin 26:88 (in Chinese)

Wang YW, Li HX, Wang GH, Tian G (1998) Breeding of foxtail millet highly sterile line "Chang10A". Gansu Agr Sci Techn 12:12–13

Wang R, Gao J, Liang GH (1999) Identification of primary trisomics and other aneuploids in foxtail millet. Plant Breed 118:59–62

Wang RQ, Gao JH, Mao LP, Du RH, Diao XM, Sun JS (2002) Chromosome location of the male-sterility and yellow seedling gene in line 1066A of foxtail millet. Acta Botanica Sin 44:1209–1212 (in Chinese)

Wang J, Wang ZL, Yang HQ, Yuan F, Guo EH, Tian G, An YH, Li HX, Wang YW, Diao XM, Guo PY (2013) Genetic analysis and preliminary mapping of a highly male-sterile gene in foxtail millet (*Setaria italica* L. Beauv.) using SSR markers. J Integr Agri 12:2143–2148

Yuan AP, Hou AB, Zhang FY, Guo YD (2008) Inheritance and effects of the photoperiod sensitivity in foxtail millet (*Setaria italica* P. Beauv). Hereditas 145:147–153

Yu-Long S, Liang-Ming W, Gai-Sheng Z, Ying S, Ya-Xin L, Zhuo-Jun Z et al (2011) Male sterility induced by chemical hybridizing agent SQ-1 in *Setaria italica* Beauv. Acta Agron Sin 9:1695

Zangre GR, Darmency H (1993) Potential for selection in the progeny of an interspecific hybrid in foxtail millet. Plant Breed 110:172–175

Zhang S, Tang C, Zhao Q, Li J, Yang L, Qie L, Fan X, Li L, Zhang N, Zhao M, Liu X, Chai Y, Zhang X, Wang H, Li Y, Li W, Zhi H, Jia G, Diao X (2014) Development of highly polymorphic simple sequence repeat markers using genome-wide microsatellite variant analysis in Foxtail millet [*Setaria italica* (L.) P. Beauv.]. BMC Genomics 15:78

Zhang H, Guo P, Wang Y, Yuan X, Dong S, Song Xe, Wang J, Wen Y (2017) Assessment of male sterility and antioxidant enzyme activities induced by the chemical hybridization agent SQ-1 in foxtail millet (*Setaria italica*). Emirates J Food Agric 29:212–221

Zhao ZH, Cui WS, Gui D, Yang S (1996) The selection of millet photo (thermo) sensitive sterile line 821 and a study on the relation of sterility to illumination and temperature. Sci Agric Sin 29:2331 (in Chinese)

Zhi H, Wang YQ, Li W, Wang YF, Li HQ, Lu P, Diao XM (2007) Development of CMS material from intra-species hybridization between green foxtail and foxtail millet. J Plant Genet Res 8:261–264

Zhu GQ, Wu QM, Ma YT (1991) Breeding of Ve type CMS in foxtail millet. Shaanxi Agric Sci 1:7 (in Chinese)

Genome-Wide Association Studies for Improving Agronomic Traits in Foxtail Millet

6

Roshan Kumar Singh and Manoj Prasad

Abstract

With the immense advancements in sequencing and data mining approaches, identification of genome-wide genetic variants in a population has become very popular. The use of these resources in the development of a dense genetic map of genome variations and to identify associated quantitative traits has become widespread in crop genetics. In recent years, genome-wide association study (GWAS) has become a powerful tool in revealing the relationship between natural variation of complex genotype and genetic locus. A slight variation in the genetic architecture of an individual in a population results in contrasting agronomic traits compared to the other individuals. GWAS utilized high-throughput genotyping platform and extensively phenotyping data to detect the links between genetic variations that underlie variations in agronomic traits. These studies can accelerate the use of genomic selection in marker-assisted breeding for crop improvement. Here, a brief discussion of available genomic resources and their utilization, quantitative trait loci (QTL) underlying agronomic traits, GWAS in foxtail millet, and the prospects for this field in crop designing is given.

6.1 Introduction

Genome-wide association study (GWAS) is a term defined by identification of a whole genome-wide set of genetic variants associated with traits within a population. With the advancements in next generation sequencing (NGS) platforms, GWAS has become the prevalent approach for interpreting the association of genotype variations with respective phenotypes. The availability of molecular markers

R.K. Singh · M. Prasad (✉)
National Institute of Plant Genome Research (NIPGR), Aruna Asaf Ali Marg, New Delhi 110067, India
e-mail: manoj_prasad@nipgr.ac.in

M. Prasad (ed.), *The Foxtail Millet Genome*, Compendium of Plant Genomes,
DOI 10.1007/978-3-319-65617-5_6

such as short sequence repeats (SSRs) and single nucleotide polymorphisms (SNPs) has been increased during the last decades that favored the detection of polymorphisms within the segment of the genome which is associated with the specific trait (Muthamilarasan et al. 2014a; Biscarini et al. 2016). Thus, GWAS mainly focuses on the association between SNPs (more prevalently) identified throughout the genome with the phenotypes such as plant disease resistance, abiotic stress tolerance, yield, plant architecture, nutritional properties, metabolomics, life cycle duration, etc. During the performance of GWAS, the causal loci need to be identified and mapped, and thereafter the responsible gene need to be identified and characterized (Ogura and Busch 2015).

Foxtail millet [*Setaria italica* (L.) P. Beauv.], considered as one of the oldest cultivated grain crops, was domesticated more than 8,700 years ago in northern China (Zohary and Hopf 2000). It is an excellent drought-tolerant crop, requires minimal water and warm weather for growth, and is mostly cultivated in dry semi-arid areas of Indian, China, the northern part of Africa, and America. It is the second most important millet crop after pearl millet in terms of global millet production. Foxtail millet has become an excellent model crop for the study of abiotic stress tolerance mechanisms in C_4 plants, grass genomics, and biomass production for biofuel crops because of its small size and true diploid genome ($\sim$515 Mb; $2n = 2x = 18$), short life cycle ($\sim$120 days), self-pollinating crop, large germplasm collection, availability of reference genome sequences, and less repetitive DNA in genome (Doust et al. 2009; Muthamilarasan and Prasad 2015, 2017). Besides, foxtail millet is closely related to many of the C_4 biofuel grasses with complex genome architecture, such as switchgrass (*Panicum virgatum*), napier grass (*Pennisetum purpureum*), and pearl millet (*Pennisetum glaucum*). Foxtail millet is nutritionally rich compared to other cereal crops. Seeds of foxtail millet contain higher proteins (14–16%), crude fat (3–7%), dietary fibers ($\sim$8%), antioxidants, and minerals, have a low glycemic index (GI), and higher value of resistant starch (Muthamilarasan et al. 2016a). Foxtail millet is therefore a suitable diet for diabetic patients, particularly those with type 2 diabetes (Thathola et al. 2010; Itagi et al. 2012; Jali et al. 2012).

There is a large number of cultivars available in foxtail millet with wide genetic diversities and phenotype variations, such as plant height, time of flowering, inflorescence trait, life cycle durations, yield, and grain phenology, and therefore characterizing these genomic resources is a necessity for the improvement of these cultivars (Reddy et al. 2006). For further advanced research, GWAS for agronomic traits has been performed in foxtail millet (Jia et al. 2013). It is also carried out in other grain crops such as rice (Huang et al. 2010, 2012; Zhao et al. 2011), maize (Kump et al. 2011; Tian et al. 2011; Li et al. 2013), and sorghum (Morris et al. 2013).

6.2 Genomic Resources in Foxtail Millet

6.2.1 Germplasm Collection of Foxtail Millet for Genetic Diversity Study

With more than 46,000 cultivated and more than 900 wild accessions of foxtail millet, there is a massive germplasm collection in gene banks throughout the world (Lata et al. 2013). Core and mini-core collections have been developed from the available germplasm sets and have served as genomic resources for genetic diversity study. China has the largest collection of foxtail millet germplasm from different countries, followed by India, France, and Japan. A core collection of 155 and 184 foxtail millet accessions has been reported at The International Crops Research Institute for the Semi-Arid Tropics (ICRISAT) gene bank (Upadhyaya et al. 2008) and National Institute of Plant Genome Research (NIPGR) (Gupta et al. 2014). All the available 155 ICRISAT foxtail millet accessions (trait-specific germplasm) have been evaluated for 21 characters at 5 diverse agro-ecological conditions and this resulted in identification of accessions with superior agronomical (high grain yield, early maturity, panicle size, shape, and seed phenology) and nutritional traits (high seed protein, iron, zinc, and calcium), which can be used in breeding programs for crop improvement (Upadhyaya et al. 2011). Similarly, the 184 accessions available in our laboratory at NIPGR have also been evaluated and for population structure determination, identification of accession with superior agronomical and nutritional traits, molecular markers development, and allele mining for essential abiotic stress-related, agronomical and nutritional traits (Lata et al. 2011a, b; Gupta et al. 2014). Genetic variability study has been carried out with 741 germplasm accessions by Nirmalakumari and Vetriventhan (2010) to analyze the yield-contributing traits in foxtail millet. Their study suggested that for improved grain yield potential, more productive tillers and tillers of medium length and medium flowering duration should be taken. It is very important for plant breeders to understand the interactions among various traits themselves and with the yield potential, which is of enormous use in crop improvement by breeding practice.

The accessions of foxtail millet germplasm have been classified into three races on the basis of inflorescence morphology, namely *moharia*, *maxima*, and *indica*. Race *moharia*, predominantly found in Europe and southwestern Asia, is characterized by cultivars with 5–52 culms, each having numerous, small, more or less erect inflorescences. Accessions in race *maxima* predominately occur in Transcaucasian Russia and the Far East, and are characterized by plants with mostly unbranched 1–8 culms with large inflorescence. The cultivars in race *indica* are cultivated mainly in southern Asia and possess intermediated number of culms (average 6.6) and inflorescence size between *moharia* and *maxima*. The three races were further subdivided into ten subraces, namely *aristata*, *fusiformis*, and *glabra* grouped in *moharia*, *compacta*, *spongiosa*, and *assamense* present in *maxima*, and *erecta*, *glabra*, *nana*, and *profusa* in *indica* (Prasada Rao et al. 1987).

6.2.2 Genome Sequence and Sequence-Based Phylogeny of Foxtail Millet

With the recent improvements in genome sequencing technologies, through the advancement of high throughput NGS techniques, it has become possible to sequence the whole genome of a species within a short period, develop millions of molecular markers throughout the genome, and understand the genetic correlation and complexity of the genome between different species. All genes identification within a species through bioinformatic analysis, its functional annotation, and molecular markers associated with the genes are pre-requisite for GWAS. Foxtail millet genome sequence has been available from two different independent studies (Bennetzen et al. 2012; Zhang et al. 2012) and serves as a reference genome for genetic studies. Foxtail millet possesses one of the smallest genomes among the panicoid family, with a genome size of ~515 Mb. Bennetzen et al. (2012) have produced the whole genome sequence of the Yugu 1 cultivar of foxtail millet through Sanger ABI3730xl platform with ~12× genomic coverage. The genomic assembly was anchored to 992-locus genetic map and consisted of a comparison between more than 1.3 million reads of expressed sequence tags. Full genome annotation revealed that ~40% genome of Yugu 1 cultivar consists of transposable elements, among which the long terminal repeat retrotransposons are predominant. A total of 24,000–29,000 protein-encoding genes were predicted, among which 10,590 were annotated as single intron genes. In another similar study, Zhang et al. (2012) used Zhang gu cultivar of foxtail millet where whole genome shotgun combined with NGS using Illumina GA II sequence platform with the genome coverage of ~10×. The final genome assembly represented 46% of the transposable elements and annotated to a total of 38,801 protein-encoding genes, of which 81% were estimated to be expressed according to the data of mRNA sequencing.

The phylogenetic relationship between grass families has been demonstrated after comparing the draft genome of foxtail millet to other members of the grass family such as rice, *Brachypodium*, maize, and sorghum. The genus *Setaria* belongs to subfamily Panicoideae and tribe Paniceae. Paniceae includes millets group, and switchgrass is closely related to tribe Andropogoneae, which consists of sorghum and maize. Foxtail millet also utilizes NADP-dependent malic enzyme subtype of C_4 photosynthetic pathways such as in sorghum and maize. *Setaria* and pearl millet share a common ancestor from ~8.3 million years ago, and *Setaria* and Panicum (switchgrass and proso millet) from ~13.1 million years ago. *Setaria* separated from sorghum and maize ~27 million years ago. It was estimated that a whole genome duplication event had occurred to all members of the grass family ~70 million years ago, before the separation of *Setaria* from maize and sorghum. There is highly conserved collinearity between genomic regions of foxtail millet and rice (71.8%), maize (86.7%), *Brachypodium* (61.5%), and sorghum (72.1%) that indicates a close evolutionary relationship between these grasses (Zhang et al. 2012).

6.3 Dissecting Genetic Diversity and Allele Mining in Foxtail Millet

On the basis of genetic diversity in the genome, molecular markers have been developed, monitored and analyzed between cultivars and across generations. The association between a molecular marker and inherited traits is used to associate the genotype of an individual with its expressed phenotype. The development of millions of novel markers associated with the agronomic traits can revolutionize plant breeding research (Edwards and Batley 2010). With the recent advance in NGS technology, the development of genomic and transcriptomic resources is very rapid and cost-effective, which can be used to develop genome-wide molecular markers such as microsatellites (SSRs) and SNPs. A genome-wide marker is often developed by comparing genome sequences from whole genome re-sequencing efforts where sequences from the genomes of several accessions were aligned to generate larger numbers of SSRs and SNPs for exploring genetic diversity within species, generating haplotype maps and executing GWAS (Kumpatla et al. 2012). In foxtail millet, initially, SSR markers were used to identify the regions of the genome associated with expressed phenotypic traits (Gupta et al. 2014). However, the major limitation of these studies was insufficient genome coverage because of fewer available SSR markers. Currently, SNP markers have been widely used in foxtail millets because of their larger number of collections and the availability of high-density genetic maps based on SNPs (Wang et al. 2010).

Microsatellite markers (SSR) show polymorphisms in a number of their repeat units and are highly significant in plant breeding because of their multiallelic nature, co-dominant inheritance, genome specificity, and genome-wide abundance (Gupta and Varshney 2000; Ganal and Roder 2007). They play an important role in gene tagging, identification, and validation of trait-associated quantitative trait loci (QTLs), physical mapping of genes and QTLs to chromosomes, and genomic selections during marker-assisted breeding (Singh et al. 2017). Jia et al. (2009) have developed 81 SSR markers and 20 RFLP markers from the genomic DNA of two mapping parents of F2 population, namely 'B100' of cultivated *S. italica* and 'A10' of its wild ancestor *S. viridis*. All the markers were mapped onto nine chromosomes of foxtail millet with total map length of 1,654 cM (centiMorgan) with an average marker density of 16.4 cM. In another study, a total of 174 highly polymorphic microsatellite markers were developed from the foxtail millet cultivars, and showed high levels of polymorphic potential (52%) and cross-species transferability (~74%) in six different grass species, namely rice (*Oryza sativa* L. 'IR64'), maize (*Zea mays* L. 'B73'), wheat (*Triticum aestivum* L. 'PH132'), pearl millet (*Pennisetum glaucum* L. 'T1'), sorghum (*Sorghum bicolor* L. Moench 'BTX623'). and guinea grass (*Panicum maximum* L. 'SPM92') (Gupta et al. 2012). A total of 534 EST-derived SSR markers have also been developed in foxtail millet with a high percentage (an average of ~88%) of cross-genera amplification (Kumari et al. 2013). In the same year a genome-wide study was carried out, resulting in the identification of 28,324 SSR motifs spanning 405.3 Mb of the genome with an average of ~69 SSR markers per Mb of foxtail millet genome (Pandey et al. 2013). They have reported that the abundance of trinucleotide repeats (48%) is higher than dinucleotide repeats (46%) in foxtail millet genome.

Because of the availability of reference genome, it is relatively easy to identify SNPs within the plant species by re-sequencing and aligning the genome sequence from an available cultivar of the species. Therefore, currently, SNPs are the most dominantly used molecular markers in plant breeding and molecular genetics. Apart from the identification of genome-wide SNPs within the plant, the detection of gene-based SNPs or SNP within ESTs are significant for GWAS (Fig. 6.1). SNPs are superior to SSRs by being more abundant in the genome and ubiquitous in the development of ultra-high throughput assays, although less polymorphic than microsatellites because of their bi-allelic nature. A total of 916 diverse varieties of foxtail millet have been sequenced by Jia et al. (2013) to determine the genetic variations within foxtail millet germplasm, and resulted in the identification of 2.58 million SNPs, among which 0.8 million common SNPs were used to construct a haplotype map of foxtail millet genome. Recently, Upadhyaya et al. (2015) have identified 17,417 SNPS from the genomic sequencing of 181 foxtail millet accessions, which were used to determine the genetic diversity of germplasm accessions and study of foxtail millet population genetics. The identified SNPs were distributed to all nine chromosomes of foxtail millet, although concentrated mainly along the subtelomeric rather than the pericentromeric region. Most of the SNPs were located in intergenic regions

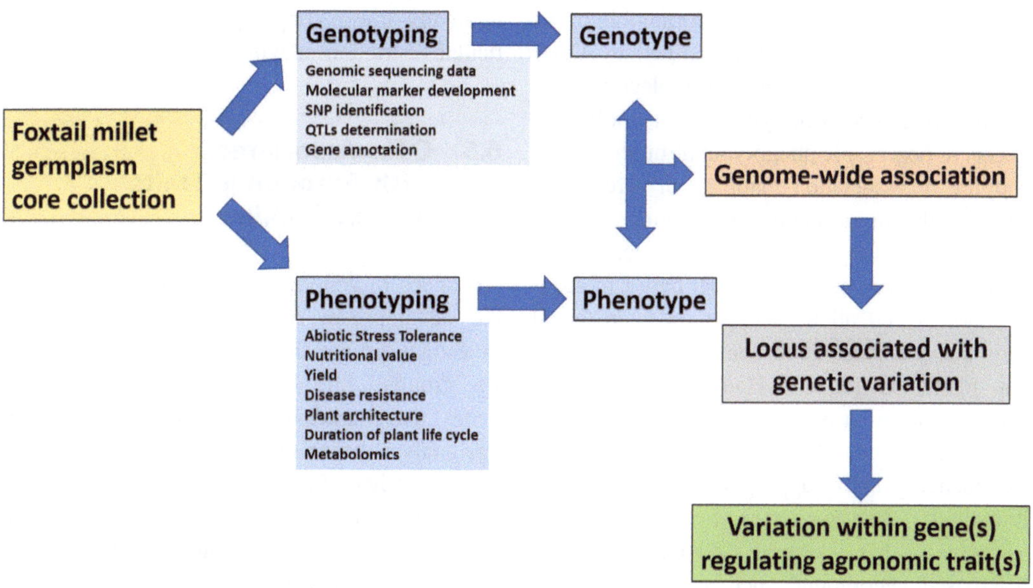

Fig. 6.1 A schematic representation of genome-wide association study (GWAS) in foxtail millet. The genomic sequence information of genetically diverse foxtail millet accessions utilized for molecular markers development and genotype calling. All accessions thoroughly phenotyped for agronomic traits and GWAS can be performed with the molecular markers and phenotyping data. Locus associated and candidate genes responsible for variation in traits can be predicted through gene annotation, expression profiling and genic variation identification

(23%), followed by exonic regions (12%), and in introns (4%). Genotype by sequencing (GBS) and GWAS has also been performed in this study, which is discussed in the succeeding part of this chapter. One fine example of SNP associated with a foxtail millet gene and controlling agronomic trait has been characterized by Lata et al. (2011a). A single synonymous SNP at a 558th base pair (A/G transition) of *SiDREB2* gene has been identified, which is associated with dehydration tolerance in a core set of 45 foxtail millet accessions (Lata et al. 2011a). An allele-specific marker (ASM) for dehydration tolerance has also been developed based on this SNP, which has been validated further in a core set of 170 accessions of foxtail millet (Lata and Prasad 2012).

6.4 High Throughput Genotyping for Validation of Molecular Markers

Genotyping of each accession can determine genomic diversity within germplasm. The genotype of an individual is the information of its complete genetic architecture, defined by DNA sequence information, allele and molecular marker patterns. The genotype of an individual reflects in observable phenotype, through linkage mapping or association mapping in a recombinant population or natural population, respectively. Before the growth of NGS technology, polymerase chain reaction (PCR)-based molecular markers and allelic scoring on agarose gel were the preferred genotyping approaches for crop plants (Huang and Han 2014). With the recent advances in NGS techniques, genotyping has mainly been based on whole genome re-sequencing and mapping of available SNPs throughout the genome. Sequencing-based genotyping has several advantages over PCR-based genotyping, such as cost effectiveness, less time consumption, and feasibility for use with crops of large genome size and high diversity. The greatest impact of high-throughput genotyping is the use of these techniques to identify genomic regions, QTLs, and causative genes in non-model crops with heritable diversity for important agronomic traits. There are a number of high-throughput genotyping methods, including microarray-based genotyping, sequencing-based genotyping, GBS, exon sequencing-based genotyping, and RNA-seq-based genotyping that are commonly used. In the case of foxtail millet, sequence-based genotyping (Jia et al. 2013) and GBS methods (Upadhyaya et al. 2015) have been successfully used. In the case of sequence-based genotyping, whole-genome re-sequencing with low coverage has been carried out for mapping population, and genome-wide SNPs were identified to construct a haplotype map of the genome (Jia et al. 2013). Another approach is to construct GBS libraries by digesting genomic DNA with restriction enzymes and ligating the digested DNA fragments to specific adaptors. This method is simple, cost-effective, less time consuming, highly reproducible, and applicable to crop species of large genome size (Xu et al. 2012). The method uses methylation-sensitive restriction enzymes, which results in avoidance of repetitive regions of the genome and enhanced accessibility of relatively lower copy regions flanking the particular restriction enzymes.

6.5 QTLs Associated with Agronomic Traits in Foxtail Millet

Advances in genomics technologies have led to the development of dense genetic maps and investigation into the relationship between linkage groups (chromosomes), molecular markers, genes, and marker-associated phenotypes. Linkage analysis-based QTL mapping is the conventional method for mapping QTLs regulating agronomic traits, and involves segregating mapping populations such as double haploids (DHs), F2 or recombinant inbred lines (RILs), polymorphic molecular markers, genetic map construction through genotyping of segregating mapping population with molecular markers, detailed phenotyping for agronomically importance traits, and mapping of QTLs utilizing both phenotyping and genotyping data (Mir et al. 2012). However, this method is not considered so accurate for mapping QTLs because of intrinsic limitations associated with every mapping population, including insufficient genome shuffling through meiotic recombination results in occurrence of identified QTLs to large genomic regions, limited variations in phenotype for the traits within mapping populations, and segregation of different QTLs associated with the same phenotype within different mapping populations (Myles et al. 2009). In recent years, linkage disequilibrium (LD)-bases association mapping is widely used as a method for QTL mapping in crop plants. Association mapping involves mapping populations consisting of hundreds of F2s or RILs representing every accession of germplasm to be precisely phenotyped and genotyped using thousands of high-density molecular markers. Association mapping populations experience a large number of past recombination events resulting in high segregation of blocks of molecular markers associated with QTLs and enhanced QTL region resolution. QTLs mapped through linkage analysis-based mapping have low QTL region resolution with an average interval of 10–20 cM. Association mapping is an approach to detect a significant relation between agronomic traits associated with the gene(s), or molecular markers which are at linkage disequilibrium (Kumpatla et al. 2012). Agronomic traits and molecular markers must be in linkage disequilibrium if their association is true and they have been transferring from generation to generation together. The higher the value of linkage disequilibrium, the greater the association between markers and QTL, but the high linkage disequilibrium interferes with the identification of genes regulating agronomic traits. There are several studies that have been reported as mapping QTLs controlling primary agronomically important traits in major cereal crops but only a limited study are available for foxtail millet.

QTLs associated with vegetative branching patterns have been identified and revealed that basal branching (tiller branching) and axillary branching are controlled by different loci in foxtail millet (Doust et al. 2004). Four QTLs (one each on linkage group I and V and two on linkage group III) were found to associate with basal branching, and four QTLs (single on each linkage group VI and IX and double on linkage group V) were identified for controlling axillary branching. Several hormonal biosynthesis pathway genes (auxin and gibberellic acid biosynthesis pathway genes) and some transcriptional regulator genes were found to be associated with the QTLs controlling basal and axillary branching (Doust et al. 2004). Mauro-Herrera et al. (2013) have utilized the SSR markers developed by Jia et al. (2009), Wang et al. (2009), and Gupta et al. () to identify the genetic loci regulating flowering in *S. italica* and its wild ancestor *S. viridis*. A total of 16 QTLs controlling 2012flowering time were identified and co-localization of these QTLs during segregation regulated differences in flowering time. A detailed comparison of QTLs for flowering time in *Setaria*, sorghum, and maize reveals that flowering time variation in separate grass lineage is regulated jointly by conserved and lineage-specific genes. In another study, 18 QTLs were identified from an interspecific mapping population generated by a cross between *S. italica* cultivar "Yugu1" X and *S. viridis* accession "W53" (Qie et al. 2014). Among 18 QTLs, 10 QTLs (3 on chromosome number 1, 2 each on chromosome numbers 6, 7, and 9, and 1 on chromosome number 5) were involved in controlling drought tolerance among the mapping populations. Three QTLs (one each on chromosome numbers 3, 7, and 9) were identified to regulate osmotic stress and five QTLs (two on chromosome number 7 and one each on chromosome numbers 5, 2, and 1) were detected under normal control conditions. The study suggested that the genotype of a wild cultivar of foxtail millet (*S. viridis*) serves as a reservoir for novel alleles associated with providing stress tolerance and could be utilized in foxtail millet improvement through breeding. Fang et al. (2016) have developed a total of 10,598 SSR markers of which 1,013 showed high polymorphism in foxtail millet accessions. Utilizing the genetic map constructed through these markers, 29 QTLs were detected for 11 agronomic and yield-related traits with 7.0–14.3% of phenotypic variations. It was found that the many QTLs associated with different phenotypes were co-located in the same intervals of the genome. The QTLs identified as associated with different agronomic traits could be used to accelerate the foxtail millet breeding program through marker-assisted selection.

6.6 Genes Associated with Agronomic Traits in Foxtail Millet

Being very tolerant to abiotic stress, most of the studies in foxtail millet have been carried out on gene families responsible for abiotic stress tolerance. A number of genes have been identified and characterized for their role in deciphering abiotic stress tolerance in this crop. In comparison with drought and salinity stress, limited studies have been reported for genes associated with plant architecture (Doust and Kellogg 2006; Preston and Kellog 2006; Liu et al. 2009; Luan et al. 2010), nutritional values (Fukunaga et al. 2002), and herbicide resistance (Wang and Darmency 1997) in foxtail millet. The first report on identification of differentially expressed transcripts through c-DNA array during salinity stress has been produced by Sreenivasulu et al. (2004). Salinity-induced phospholipid hydroperoxide glutathione peroxidase (*PHGPX*) gene has been characterized and it has been suggested that it plays an important role in conferring tolerance to salinity-induced oxidative damage. Similarly, Zhang et al. (2007a, b) have constructed the EST library of dehydration stressed foxtail millet cultivar MAR51 and identified 95 ESTs upregulated in roots and 57 in shoots. The differentially expressed transcripts from roots were different from the upregulated transcripts reported from the shoots, and most of the upregulated genes were found to be associated with protein degradation pathway. They have further characterized a putative 12-oxophytodienoic acid reductase 1 (*SiORP1*) gene which upregulated in roots during osmotic stress and were unaffected by ABA (abscisic acid), NaCl, and MeJA (methyl jasmonate) treatments and hence might play a significant role in drought stress response (Zhang et al. 2007a, b). Veeranagamallaiah et al. (2007, 2009) has reported that, with the onset of salinity stress, the concentration of molecules such as proline and sorbitol increases in tolerant cultivar of foxtail millet. The genes responsible for the accumulation of proline and sorbitol, namely glutamine synthetase and pyrroline-5-carboxylate reductase for proline and aldose reductase for sorbitol are highly upregulated in tolerant cultivar of foxtail millet compared to the susceptible one. Comparative transcriptome analysis of foxtail millet during dehydration stress has been performed by Lata et al. (2010), resulting in the identification of 86 upregulated genes among 327 unique ESTs through the reverse northern method. The expression pattern has been further validated by quantitative real-time PCR which showed $\geq$2.5-fold induction of genes after dehydration stress. A dehydration responsive element-binding-type (DREB2) protein was shown to be upregulated up to 11-fold which was further cloned and characterized for its role in dehydration stress in foxtail millet (Lata et al. 2010, 2011a, b, 2013; Lata and Prasad 2012). Li et al. (2014) have characterized an ABA-responsive DREB-binding protein (*SiARDP*) gene which shows high upregulation during drought, salinity, and cold stress as well as upon external ABA treatment to foxtail millet seedling. The overexpression of *SiARDP* in Arabidopsis enhances tolerance to drought and salt stress during the germination and seedling development stages (Li et al. 2014). A membrane-bound *NAC* gene (*SiNAC*) has been identified by Puranik et al. (2011a) from the SSH library of foxtail millet constructed after comparison of the transcriptome of salinity tolerant and susceptible cultivar. Cloning and molecular characterization of the *SiNAC* gene revealed that this particular gene has a DNA-binding site and regulates the transcription of many stress-responsive genes as well as many development associated genes (Puranik et al. 2011b, c). The role of the NAC gene family is not only restricted to abiotic stress; it also plays an important role in plant development, biotic stress, cell cycle control, secondary wall development, and senescence (Puranik et al. 2013).

In another study, Mishra et al. (2014) identified a *WD40* gene expressing differentially in dehydration and salinity stress in foxtail millet. Further characterization of the *SiWD40* gene showed that it acts as a scaffolding molecule in diverse abiotic stress-induced protein-protein interactions to meet their functions (Mishra et al. 2012). The whole transcriptome of foxtail millet has been analyzed by Qi et al. (2013) through next generation deep sequencing techniques and they identified 2,824 genes affected by drought stress of which expression of 48.2% was upregulated and 51.8% genes were downregulated. Late embryogenesis abundant protein, heat shock proteins (HSPs), dehydrin, aquaporin, and phosphatase 2C were predominant among the upregulated genes, giving an indication of the possible role of these genes in dehydration response in crop plants. Singh et al. (2016) has performed a genome-wide study of HSP genes in foxtail millet and identified a total of 113 putative HSPs of which 37, 20, 27, 9, and 20 belongs to *sHSP*, *HSP60*, *HSP70*, *HSP90*, and *HSP100*, respectively, based on their molecular weight. Apart from heat stress, many of the HSPs from each class have shown differential expression during salinity, dehydration and cold stress. *SisHSP27*, a small molecular weight heat shock protein, was identified as a potential candidate gene which showed substantial higher expression in heat stress and moderate upregulation in salinity, dehydration, and low temperature in tolerant cultivar. Functional characterization by overexpression of *SisHSP27* gene in *Saccharomyces cerevisiae* confers tolerance to heat, salinity, and dehydration stress, and thus it can be concluded that this *SisHSP27* might play a significant role in providing tolerance to abiotic stress (Singh et al. 2016). Similarly, other gene families such as ADP-ribosylation factors (Muthamilarasan et al. 2016b), WRKY (Muthamilarasan et al. 2015a), secondary cell wall synthesis genes (Muthamilarasan et al. 2015b), MYB (Muthamilarasan et al. 2014b), and C_2H_2-type of zinc finger transcription factors (Muthamilarasan et al. 2014c) were also characterized in foxtail millet. All the identified and functionally characterized genes from foxtail millet can be exploited as an important resource for crop improvement using direct transfer to foxtail millet by genetic transformation or by plant breeding methods.

6.7 Conclusions and Future Perspectives

Recent advances in genome sequencing technology, accessibility of high throughput genotyping platforms, and the capabilities of data processing have led to widespread utilization of GWAS to identify polymorphisms in DNA sequences responsible for variation in agronomic traits in genetically distinct mapping populations. With the advent of densely distributed genome-wide molecular markers as SSRs and preferably SNPs typing, GWAS aids in the identification of phenotypic trait-contributing loci at high resolution and genes associated with the expressed phenotypes. Mapping populations used for GWAS are of diverse genetic backgrounds, representing each available accession or RILs with ancient recombination events that has occurred in the genome to identify the loci-controlling phenotype with high resolution. The method of GWAS has been extensively used during the last decade in human genetics, especially to identify genes associated with human diseases. With the recent availability of developed genomic resources, GWAS has been successfully carried on many crop species including foxtail millet. For the first time, GWAS has been performed on foxtail millet by Jia et al. (2013) in which a total of 916 accessions, including both modern cultivars and traditional landraces, were finely genotyped through low-coverage ($\sim 0.7\times$) whole genome sequencing methods. Extensive phenotyping of all the 916 accessions was performed under 5 different environmental conditions and 512 loci identified associated with 47 agronomic traits, including flowering time, plant architecture, and grain yield. The genome-wide LD decay rate in foxtail millet was ~ 100 kb on average, which is similar to that of cultivated rice

(Huang et al. 2010, 2012) and other self-fertilizing species.

Recently, Upadhyaya et al. (2015) genotyped a set of 190 foxtail millet germplasm accessions through GBS, and GWAS for flowering time and pigmentation has been performed. Genomic regions and SNP loci associated with these traits were identified and reported co-localization of putative candidate genes with associated SNPs. Taken together, the results from both the GWAS studies in foxtail millet could be further utilized to identify the genetic basis of plant architecture, physiology, and yield in multiple environments and applied in genetic improvement through genomic selection during marker-assisted breeding. The major advantage of GWAS is the use of germplasm accessions with high levels of genetic diversity which results in extensive phenotyping. On the basis of present study, it seems that GWAS is successfully achieving the goal of associating genetic differences with corresponding variations in phenotype in given populations. The increase in the number of GWAS in crops such as foxtail millet should result in the identification of single, well-annotated causative genes and clarify the diversity of genome structure associated with the individually agronomic trait of interest.

Acknowledgements Studies on millet genomics in Dr. Manoj Prasad's laboratory are supported by Science and Engineering Research Board (SERB), Department of Science and Technology (DST), Government of India [Grant No. EMR/2015/000464], by Department of Biotechnology, Government of India [Grant No. BT/HRD/NBA/37/01/2014], and by Core Grant of National Institute of Plant Genome Research (NIPGR), New Delhi, India. Roshan K. Singh acknowledges the research fellowship received from Council of Scientific and Industrial Research, Govt. of India.

References

Bennetzen JL, Schmutz J, Wang H, Percifield R, Hawkins J, Pontaroli AC, Estep M, Feng L, Vaughn JN, Grimwood J, Jenkins J, Barry K, Lindquist E, Hellsten U, Deshpande S, Wang X, Wu X, Mitros T, Triplett J, Yang X, Ye CY, Mauro-Herrera M, Wang L, Li P, Sharma M, Sharma R, Ronald PC, Panaud O, Kellogg EA, Brutnell TP, Doust AN, Tuskan GA, Rokhsar D, Devos KM (2012) Reference genome sequence of the model plant Setaria. Nat Biotechnol 30:555–561

Biscarini F, Cozzi P, Casella L, Riccardi P, Vattari A, Orasen G, Perrini R, Tacconi G, Tondelli A, Biselli C, Cattivelli L, Spindel J, McCouch S, Abbruscato P, Valé G, Piffanelli P, Greco R (2016) Genome-wide association study for traits related to plant and grain morphology, and root architecture in temperate rice accessions. PLoS ONE 11:e0155425

Doust AN, Kellogg EA (2006) Effect of genotype and environment on branching in weedy green millet (*Setaria viridis*) and domesticated foxtail millet (*Setaria italica*) (Poaceae). Mol Ecol 15:1335–1349

Doust AN, Devos KM, Gadberry MD, Gale MD, Kellogg EA (2004) Genetic control of branching in foxtail millet. Proc Natl Acad Sci USA 101:9045–9050

Doust AN, Kellogg EA, Devos KM, Bennetzen JL (2009) Foxtail millet: a sequence-driven grass model system. Plant Physiol 149:137–141

Edwards D, Batley J (2010) Plant genome sequencing: applications for crop improvement. Plant Biotechnol J 8:2–9

Fang X, Dong K, Wang X, Liu T, He J, Ren R, Zhang L, Liu R, Liu X, Li M, Huang M, Zhang Z, Yang T (2016) A high density genetic map and QTL for agronomic and yield traits in foxtail millet [*Setaria italica* (L.) P. Beauv]. BMC Genom 17:336

Fukunaga K, Kawase M, Kato K (2002) Structural variation in the waxy gene and differentiation in foxtail millet [*Setaria italica* (L.) P. Beauv.]: implications for multiple origins of the waxy phenotype. Mol Genet Genomics 268:214–222

Ganal MW, Roder MS (2007) Microsatellite and SNP markers in wheat breeding. In: Varshney RK, Tuberosa R (eds) Genomic assisted crop improvement: genomics applications in crops, vol 2. Springer, Dordrecht, pp 1–24

Gupta PK, Varshney RK (2000) The development and use of microsatellite markers for genetic analysis and plant breeding with emphasis on bread wheat. Euphytica 113:163–185

Gupta S, Kumari K, Sahu PP, Vidapu S, Prasad M (2012) Sequence based novel genomic microsatellite markers for robust genotyping purposes in foxtail millet [*Setaria italica* (L.) P. Beauv.]. Plant Cell Rep 31:323–337

Gupta S, Kumari K, Muthamilarasan M, Parida SK, Prasad M (2014) Population structure and association mapping of yield contributing agronomic traits in foxtail millet. Plant Cell Rep 33:881–893

Huang X, Han B (2014) Natural variations and genome-wide association studies in crop plants. Annu Rev Plant Biol 65:531–551

Huang X, Wei X, Sang T, Zhao Q, Feng Q, Zhao Y, Li C, Zhu C, Lu T, Zhang Z, Li M, Fan D, Guo Y, Wang A, Wang L, Deng L, Li W, Lu Y, Weng Q, Liu K, Huang T, Zhou T, Jing Y, Li W, Lin Z, Buckler ES, Qian Q, Zhang QF, Li J, Han B (2010) Genome-wide association studies of 14 agronomic traits in rice landraces. Nat Genet 42:961–967

Huang X, Zhao Y, Wei X, Li C, Wang A, Zhao Q, Li W, Guo Y, Deng L, Zhu C, Fan D, Lu Y, Weng Q, Liu K, Zhou T, Jing Y, Si L, Dong G, Huang T, Lu T, Feng Q, Qian Q, Li J, Han B (2012) Genome-wide association study of flowering time and grain yield traits in a worldwide collection of rice germplasm. Nat Genet 44:32–39

Itagi S, Naik R, Bharati P, Sharma P (2012) Readymade foxtail millet mix for diabetics. Int J Sci Nat 3:131–134

Jali M, Kamatar M, Jali S, Hiremath M, Naik R (2012) Efficacy of value added foxtail millet therapeutic food in the management of diabetes and dyslipidamea in type 2 diabetic patients. Recent Res Sci Tech 4:7

Jia X, Zhang Z, Liu Y, Zhang C, Shi Y, Song Y, Wang T, Li Y (2009) Development and genetic mapping of SSR markers in foxtail millet [*Setaria italica* (L.) P. Beauv.]. Theor Appl Genet 118:821–829

Jia G, Huang X, Zhi H, Zhao Y, Zhao Q, Li W, Chai Y, Yang L, Liu K, Lu H, Zhu C, Lu Y, Zhou C, Fan D, Weng Q, Guo Y, Huang T, Zhang L, Lu T, Feng Q, Hao H, Liu H, Lu P, Zhang N, Li Y, Guo E, Wang S, Wang S, Liu J, Zhang W, Chen G, Zhang B, Li W, Wang Y, Li H, Zhao B, Li J, Diao X, Han B (2013) A haplotype map of genomic variations and genome-wide association studies of agronomic traits in foxtail millet (*Setaria italica*). Nat Genet 45:957–961

Kumari K, Muthamilarasan M, Misra G, Gupta S, Subramanian A, Parida SK, Chattopadhyay D, Prasad M (2013) Development of eSSR-markers in *Setaria italica* and their applicability in studying genetic diversity, cross-transferability and comparative mapping in millet and non-millet species. PLoS ONE 8:e67742

Kump KL, Bradbury PJ, Wisser RJ, Buckler ES, Belcher AR, Oropeza-Rosas MA, Zwonitzer JC, Kresovich S, McMullen MD, Ware D, Balint-Kurti PJ, Holland JB (2011) Genome-wide association study of quantitative resistance to southern leaf blight in the maize nested association mapping population. Nat Genet 43:163–168

Kumpatla SP, Buyyarapu R, Abdurakhmonov IY, Mammadov JA (2012) Genomics-assisted plant breeding in the 21st century: technological advances and progress. Plant breeding. Ibrokhim Y Abdurakhmonov (ed) In tech, DOI:10.5772/37458

Lata C, Prasad M (2012) Validation of an allele-specific marker associated with dehydration stress tolerance in a core set of foxtail millet accessions. Plant Breed. doi:10.1111/j.1439-0523.2012.01983.x

Lata C, Sahu PP, Prasad M (2010) Comparative transcriptome analysis of differentially expressed genes in foxtail millet (*Setaria italica* L.) during dehydration stress. Biochem Biophys Res Commun 393:720–727

Lata C, Bhutty S, Bahadur RP, Majee M, Prasad M (2011a) Association of an SNP in a novel DREB2-like gene SiDREB2 with stress tolerance in foxtail millet [*Setaria italica* (L.)]. J Exp Bot 62:3387–3401

Lata C, Jha S, Dixit V, Sreenivasulu N, Prasad M (2011b) Differential antioxidative responses to dehydration-induced oxidative stress in core set of foxtail millet cultivars [*Setaria italica* (L.)]. Protoplasma 248:817–828

Lata C, Gupta S, Prasad M (2013) Foxtail millet: a model crop for genetic and genomic studies in bioenergy grasses. Crit Rev Biotechnol 33:328–343

Li C, Yue J, Yu J (2014) An ABA responsive DRE-binding protein gene from *Setaria italica*, SiARDP, the target gene of SiAREB, played critical role under drought stress. In: International Setaria Genetics Conference, p 62

Li H, Peng Z, Yang X, Wang W, Fu J, Wang J, Han Y, Chai Y, Guo T, Yang N, Liu J, Warburton ML, Cheng Y, Hao X, Zhang P, Zhao J, Liu Y, Wang G, Li J, Yan J (2013) Genome-wide association study dissects the genetic architecture of oil biosynthesis in maize kernels. Nat Genet 45:43–50

Liu Y, Feng X, Xu Y, Yu J, Ao G, Peng Z, Zhao Q (2009) Overexpression of millet ZIP-like gene (SiPf40) affects lateral bud outgrowth in tobacco and millet. Plant Physiol Biochem 47:1051–1060

Luan Y, Wang B, Zhao Q, Ao G, Yu J (2010) Ectopic expression of foxtail millet zip-like gene, SiPf40, in transgenic rice plants causes a pleiotropic phenotype affecting tillering, vascular distribution and root development. Sci China Life Sci 53:1450–1458

Mauro-Herrera M, Wang X, Barbier H, Brutnell TP, Devos KM, Doust AN (2013) Genetic control and comparative genomic analysis of flowering time in Set*ar*ia (Poaceae). G3 (Bethesda) 3:283–295

Mir RR, Zaman-Allah M, Sreenivasulu N, Trethowan R, Varshney RK (2012) Integrated genomics, physiology and breeding approaches for improving drought tolerance in crops. Theor Appl Genet 125:625–645

Mishra AK, Puranik S, Bahadur RP, Prasad M (2012) The DNA binding activity of an AP2 protein is involved in transcriptional regulation of a stress-responsive gene, SiWD40, in foxtail millet. Genomics 100:252–263

Mishra AK, Muthamilarasan M, Khan Y, Parida SK, Prasad M (2014) Genome-wide investigation and expression analyses of WD40 protein family in the model plant foxtail millet (*Setaria italica* L.). PLoS ONE 9:e86852

Morris GP, Rhodes DH, Brenton Z, Ramu P, Thayil VM, Deshpande S, Hash CT, Acharya C, Mitchell SE, Buckler ES, Yu JM, Kresovich S (2013) Dissecting genome-wide association signals for loss-of-function phenotypes in sorghum flavonoid pigmentation traits. G3 (Bethesda) 3:2085–2094

Muthamilarasan M, Prasad M (2015) Advances in *Setaria* genomics for genetic improvement of cereals and bioenergy grasses. Theor Appl Genet 128:1–14

Muthamilarasan M, Prasad M (2017) Genetic determinants of drought stress tolerance in *Setaria*. In: Doust A, Diao X (eds) Genetics and genomics of Setaria. Springer, pp 267–289

Muthamilarasan M, Suresh BV, Pandey G, Kumari K, Parida SK, Prasad M (2014a) Development of 5123 intron-length polymorphic markers for large-scale genotyping applications in foxtail millet. DNA Res 21:41–52

Muthamilarasan M, Khandelwal R, Yadav CB, Bonthala VS, Khan Y, Prasad M (2014b) Identification and molecular characterization of MYB transcription factor superfamily in C_4 model plant foxtail millet (*Setaria italica* L.). PLoS ONE 9:e109920

Muthamilarasan M, Bonthala VS, Mishra AK, Khandelwal R, Khan Y, Roy R, Prasad M (2014c) C_2H_2-type of zinc finger transcription factors in foxtail millet define response to abiotic stresses. Funct Integr Genom 14:531–554

Muthamilarasan M, Bonthala VS, Khandelwal R, Jaishakar J, Shweta S, Nawaz K, Prasad M (2015a) Global analysis of WRKY transcription factor superfamily in *Setaria* identifies potential candidates involved in abiotic stress signaling. Front Plant Sci 6:910

Muthamilarasan M, Khan Y, Jaishankar J, Shweta S, Lata C, Prasad M (2015b) Integrative analysis and expression profiling of secondary cell wall genes in C_4 biofuel model *Setaria italica* reveals targets for lignocellulose bioengineering. Front Plant Sci 6:965

Muthamilarasan M, Dhaka A, Yadav R, Prasad M (2016a) Exploration of millet models for developing nutrient rich graminaceous crops. Plant Sci 242:89–97

Muthamilarasan M, Mangu VR, Zandkarimi H, Prasad M, Baisakh N (2016b) Structure, organization and evolution of ADP-ribosylation factors in rice and foxtail millet, and their expression in rice. Sci Rep 6:24008

Myles S, Peiffer J, Brown PJ, Ersoz ES, Zhang Z, Costich DE, Buckler ES (2009) Association mapping: critical considerations shift from genotyping to experimental design. Plant Cell 21:2194–2202

Nirmalakumari A, Vetriventhan M (2010) Characterization of foxtail millet germplasm collections for yield contributing traits. Elect J Plant Breed 1:140–147

Ogura T, Busch W (2015) From phenotypes to causal sequences: using genome wide association studies to dissect the sequence basis for variation of plant development. Curr Opin Plant Biol 23:98–108

Pandey G, Misra G, Kumari K, Gupta S, Parida SK, Chattopadhyay D, Prasad M (2013) Genome-wide development and use of microsatellite markers for large–scale genotyping applications in foxtail millet [*Setaria italica* (L.)]. DNA Res 20:197–207

Prasada Rao KE, de Wet JMJ, Brink DK, Mengesha MH (1987) Intraspecific variation and systematics of cultivated *Setaria italica*, foxtail millet (Poaceae). Econ Bot 41:108–116

Preston JC, Kellogg EA (2006) Reconstructing the evolutionary history of paralogous APETALA1/FRUITFULL-like genes in grasses (Poaceae). Genetics 174:421–437

Puranik S, Jha S, Srivastava PS, Sreenivasulu N, Prasad M (2011a) Comparative transcriptome analysis of contrasting foxtail millet cultivars in response to short-term salinity stress. J Plant Physiol 168:280–287

Puranik S, Bahadur RP, Srivastava PS, Prasad M (2011b) Molecular cloning and characterization of a membrane associated NAC family gene, SiNAC from foxtail millet [*Setaria italica* (L.) P. Beauv.]. Mol Biotechnol 49:138–150

Puranik S, Kumar K, Srivastava PS, Prasad M (2011c) Electrophoretic mobility shift assay reveals a novel recognition sequence for *Setaria italica* NAC protein. Plant Signal Behav 6:1588–1590

Puranik S, Sahu PP, Mandal SN, Venkata Suresh B, Parida SK, Prasad M (2013) Comprehensive genome-wide survey, genomic constitution and expression profiling of the NAC transcription factor family in Foxtail Millet (*Setaria italica* L.). PLoS ONE 8:e64594

Qi X, Xie S, Liu Y, Yi F, Yu J (2013) Genome-wide annotation of genes and noncoding RNAs of foxtail millet in response to simulated drought stress by deep sequencing. Plant Mol Biol 83:459–473

Qie L, Jia G, Zhang W, Schnable J, Shang Z, Li W, Liu B, Li M, Chai Y, Zhi H, Diao X (2014) Mapping of quantitative trait locus (QTLs) that contribute to germination and early seedling drought tolerance in the interspecific cross *Setaria italica* × *Setaria viridis*. PLoS ONE 9(7):e101868. doi:10.1371/journal.pone.0101868

Reddy V, Upadhyaya HD, Gowda C (2006) Characterization of world's foxtail millet germplasm collections for morphological traits. J SAT Agric Res 2:1–3

Singh RK, Jaishankar J, Muthamilarasan M, Shweta S, Dangi A, Prasad M (2016) Genome-wide analysis of heat shock proteins in C_4 model, foxtail millet identifies potential candidates for crop improvement under abiotic stress. Sci Rep 6:32641

Singh RK, Sahu PP, Muthamilarasan M, Dhaka A, Prasad M (2017) Genomics-assisted breeding for improving stress tolerance of graminaceous crops to biotic and abiotic stresses: progress and orospects. Senthil-Kumar M (ed) Plant tolerance to individual and concurrent stresses. DOI 10.1007/978-81-322-3706-8_5

Sreenivasulu N, Miranda M, Prakash HS, Wobus U, Weschke W (2004) Transcriptome changes in foxtail millet genotypes at high salinity: identification and characterization of a PHGPX gene specifically upregulated by NaCl in a salt-tolerant line. J Plant Physiol 161:467–477

Thathola A, Srivastava S, Singh G (2010) Effect of foxtail millet (*Setaria italica*) supplementation on serum glucose, serum lipids and glycosylated hemoglobin in type 2 diabetics. Diabetologia Croatica 40:23–28

Tian F, Bradbury PJ, Brown PJ, Hung H, Sun Q, Flint-Garcia S, Rocheford TR, McMullen MD, Holland JB, Buckler ES (2011) Genome-wide association study of leaf architecture in the maize nested association mapping population. Nat Genet 43:159–162

Upadhyaya HD, Pundir RPS, Gowda CLL, Reddy VG, Singh S (2008) Establishing a core collection of foxtail millet to enhance the utilization of the germplasm of an underutilized crop. Plant Genet Resour 7:177–184

Upadhyaya HD, Ravishankar CR, Narasimhudu Y, Sarma NDRK, Singh SK, Varshney SK, Reddy VG, Singh S, Parzies SK, Dwivedi SL, Nadaf HL, Sahrawat KL, Gowda CLL (2011) Identification of trait-specific germplasm and developing a mini core collection for efficient use of foxtail millet genetic resources in crop improvement. Field Crops Res 124:459–467

Upadhyaya H, Vetriventhan M, Deshpande SP, Sivasubramani S, Wallace JG, Buckler ES, Hash CT, Ramu P (2015) Population genetics and structure of a global foxtail millet germplasm collection. Plant Genome 8:1–13

Veeranagamallaiah G, Chandraobulreddy P, Jyothsnakumari G, Sudhakar C (2007) Glutamine synthetase expression and pyrroline-5-carboxylate reductase activity influence proline accumulation in two cultivars of foxtail millet (*Setaria italica* L.) with differential salt sensitivity. Environ Expt Bot 60:239–244

Veeranagamallaiah G, Jyothsnakumari G, Thippeswamy M, Reddy PCO, Surabhi G-K, Sriranganayakulu G, Mahesh Y, Rajasekhar B, Madhurarekha C, Sudhakar C (2009) Proteomic analyses of salt responses in foxtail millet (*Setaria italica* L. cv. Prasad) seedlings. Plant Sci 175:631–641

Wang T, Darmency H (1997) Inheritance of sethoxydim resistance in foxtail millet, *Setaria italica* (L.) Beauv. Euphytica 94:69–73

Wang JW, Czech B, Weigel D (2009) miR156-regulated SPL transcription factors define an endogenous flowering pathway in *Arabidopsis thaliana*. Cell 138:738–749

Wang C, Chen J, Zhi H, Yang L, Li W, Wang Y, Li H, Zhao B, Chen M, Diao X (2010) Population genetics of foxtail millet and its wild ancestor. BMC Genet 11:90

Xu Y, Lu Y, Xie C, Gao S, Wan J, Prasanna BM (2012) Whole-genome strategies for marker-assisted plant breeding. Mol Breed 29:833–854

Zhang J, Liu T, Fu J, Zhu Y, Jia J, Zheng J, Zhao Y, Zhang Y, Wang G (2007a) Construction and application of EST library from *Setaria italica* in response to dehydration stress. Genomics 90:121–131

Zhang JP, Liu TS, Zheng J, Jin Z, Zhu Y, Guo JF, Wang GY (2007b) Cloning and characterization of a putative 12-oxophytodienoic acid reductase cDNA induced by osmotic stress in roots of foxtail millet. DNA Seq 18:138–144

Zhang G, Liu X, Quan Z, Cheng S, Xu X, Pan S, Xie M, Zeng P, Yue Z, Wang W, Tao Y, Bian C, Han C, Xia Q, Peng X, Cao R, Yang X, Zhan D, Hu J, Zhang Y, Li H, Li H, Li N, Wang J, Wang C, Wang R, Guo T, Cai Y, Liu C, Xiang H, Shi Q, Huang P, Chen Q, Li Y, Wang J, Zhao Z, Wang J (2012) Genome sequence of foxtail millet (*Setaria italica*) provides insights into grass evolution and biofuel potential. Nat Biotechnol 30:549–554

Zhao K, Tung CW, Eizenga GC, Wright MH, Ali ML, Price AH, Norton GJ, Islam MR, Reynolds A, Mezey J, McClung AM, Bustamante CD, McCouch SR (2011) Genome-wide association mapping reveals a rich genetic architecture of complex traits in *Oryza sativa*. Nat Commun 2:467

Zohary D, Hopf M (2000) Domestication of plants in the old world: the origin and spread of cultivated plants in west Asia, Europe, and the Nile Valley. Third edn. Oxford University Press

Genetic Structure of Foxtail Millet Landraces

7

Kenji Fukunaga

Abstract

Studies on the genetic structure of foxtail millet landraces are reviewed. Several genetic works based on intraspecific hybrid pollen semi-sterility, isozymes, ribosomal DNA (rDNA) RFLP, nuclear RFLP, mitochondrial DNA (mtDNA) RFLP, RAPD, AFLP, transposon display (TD) markers, and single nucleotide polymorphisms (SNPs) were carried out to investigate genetic structure of foxtail millet accessions mainly from Eurasia. Most of the works suggested that China is the center of diversity of foxtail millet and landraces were differentiated in local geographical groups.

7.1 Hypotheses on Geographical Origin of Foxtail Millet

Foxtail millet, *Setaria italica* (L.) P. Beauv. is one of the oldest domesticated cereals in the Old World. When Hunt et al. (2008) reviewed archaeological sites in Eurasia, archaeological remains of foxtail millet were found at the sites of Peiligang and Cishan near the Yellow River, dating back to ca. 5,000–6,000 B.C. (Li and Wu 1996) and in prehistoric sites in Europe. Recently, more detailed studies on dispersal of cultivated plants including foxtail millet were also carried out based on archaeological data (Stevens et al. 2016). Foxtail millet has been utilized in various ways peculiar to each area of Eurasia (Sakamoto 1987), and it thought to have played an important role in early agriculture in the Old World.

Cytological studies indicated that the wild ancestor of foxtail millet is green foxtail (*S. italica* ssp. *viridis* = *S. viridis*) (Kihara and Kishimoto 1942; Li et al. 1945) and this hypothesis was also supported by further genetic data such as isozymes (Wang et al. 1995) and DNA markers (Le Thierry d'Ennequin et al. 2000). However, the geographical origin of domesticated foxtail millet cannot be determined from the distribution of ssp. *viridis*, as it is commonly found in various areas in Europe and Asia and also currently in the New World. The geographical origin of foxtail millet is, therefore, still a controversial issue.

K. Fukunaga (✉)
Prefectural University of Hiroshima, Shobara 727-0023, Japan
e-mail: fukunaga@pu-hiroshima.ac.jp

M. Prasad (ed.), *The Foxtail Millet Genome*, Compendium of Plant Genomes,
DOI 10.1007/978-3-319-65617-5_7

Vavilov (1926) stated that the principal center of diversity for foxtail millet is East Asia, including China and Japan. Harlan (1975) suggested independent domestication in China and Europe based on archaeological evidence. The archaeological, isozyme, and morphological evidence (de Wet et al. 1979; Jusuf and Pernes 1985; Li et al. 1995a, b) suggested that China is the center of diversity and original home of foxtail millet, but independent origin of this millet in other regions cannot be excluded. Furthermore, Li et al. (1995b) stated that landraces in Afghanistan and Lebanon had been domesticated independently in relatively recent times because they had primitive morphological characters such as several tillers with small panicles and looked similar to ssp. *viridis* but with non-shattering large grains. Recent archaeological evidence also supports the domestication of foxtail millet in China (Nasu et al. 2007; Hunt et al. 2008). In contrast to the hypothesis of Chinese origin and multiple origins, Sakamoto (1987) suggested that foxtail millet originated somewhere in Central Asia-Afghanistan-Pakistan-India because accessions with less compatibility (Kawase and Sakamoto 1987) and with primitive morphological traits are found there. This hypothesis, which excludes China as a center of origin of foxtail millet, is very different from the others. In this chapter we review studies on the genetic structure of foxtail millet landraces from various parts of Europe and Asia, mainly in DNA markers.

7.2 Genetic Differentiation of Foxtail Millet Landraces, Revealed by Biochemical and Genetic Markers and Intraspecific Hybrid Pollen Sterility

Several studies have been carried out to clarify the genetic structure of foxtail millet from Europe and Asia (and partly from Africa) such as biochemical markers (isozymes and prolamin) and intraspecific hybrid pollen sterility before DNA markers [nuclear RFLP, mitochondrial RFLP, RAPD, AFLP and transposon display (TD) markers and single nucleotide polymorphisms (SNPs)] became available [RFLP = restriction fragment length polymorphism, RAPD = random amplified polymorphic DNA, AFLP = amplified fragment length polymorphism]. Genetic studies on biochemical markers and intraspecific hybrid pollen sterility are discussed below.

7.2.1 Variation in Biochemical Markers (Isozymes and Prolamin)

Kawase and Sakamoto (1984) investigated polymorphism in two loci, *Est-1* and *Est-2*, of the esterase isozymes of 432 accessions of foxtail millet collected from different areas throughout Eurasia by gel isoelectric focusing. On locus *Est-1*, most of the accessions had *Est-1 a*, which was widely distributed throughout Eurasia, and 9% of accessions had *Est-1 b*, which was distributed in China and Korea. On locus *Est-2*, most of the accessions had *Est-2 a*, but nine had *Est-2 b*, which is found in all of the accessions from the western part of Europe and in one of the Indian accessions. Six had the *Est-2 c* allele, which was found in Japan and China. They concluded that the distribution of *Est-2 a* and *Est-2 b* might indicate some degree of phylogenetic differentiation between the Asian and the European accessions and that Chinese accessions showed polymorphism in both loci. Jusuf and Pernes (1985) investigated the genetic diversity of a world collection of foxtail millet accessions and some samples of wild populations (ssp. *viridis*) by electrophoresis on five enzymes (10 loci) *Est*, *Acph*, *Got*, *Mdh*, and *Pgd*. They found some genetic groups of foxtail millet in China-Korea-Japan, Okinawa (Nansei Islands of Japan)-Taiwan, India-Kenya, and Europe. They also investigated wild populations collected in France and China and concluded that it is possible that there were independent domestications in both Europe and China because foxtail millet and *S. viridis* accessions were more closely related in isozyme alleles in Europe and China.

Nakayama et al. (1999) investigated allelic variation at the two prolamin loci (*Pro1* and *Pro2*), and their geographical distribution in 560 local cultivars of foxtail millet collected mainly from Eurasia and studied using SDS-polyacrylamide gel electrophoresis (SDS-PAGE). Two alleles (*Pro1a* and *Pro1null*) at the *Pro1* locus and six alleles (*Pro2a*, *Pro2b*, *Pro2c*, *Pro2d*, *Pro2e*, and *Pro2f*) at the *Pro2* locus were detected among the cultivars examined. No apparent trend in *Pro1* was observed in geographical distribution. In contrast, two common alleles at the *Pro2* locus, *Pro2b* and *Pro2f*, had clear differential geographical distribution. The *Pro2b* allele was most frequent in Europe and decreased in frequency going eastward. The *Pro2f* allele frequently occurred in subtropical and tropical regions, including the Nansei Islands of Japan, the Philippines, Nepal, India, Pakistan, and Africa. All eight alleles at the *Pro1* and *Pro2* loci occurred in China, suggesting that China is a center of diversity. They also found a "tropical group" characterized by the *Pro2f* allele and other genes.

7.2.2 Classification by Means of Intraspecific Hybrid Pollen Sterility

Intraspecific hybrid pollen sterility can be a genetic indicator of differences. In rice, classification based on hybrid sterility was carried out (Kato et al. 1928), and Asian rice varieties were classified into two main groups, japonica and indica types. Kawase and Sakamoto (1987) crossed 83 accessions of *Setaria italica* collected from various areas throughout Eurasia with 3 tester strains from Japan (tester A), Lan Hsü Island of Taiwan (B), and Belgium (C). The accessions could be clearly classified into six types, designated as types A, B, C, AC, BC, and X. They regarded pollen fertility of more than 75% as normal. The accessions of types A, B, and C were those that produced F1 hybrids having normal pollen fertility when crossed with testers A, B, and C, respectively. When both F1 hybrids from the crosses with two testers, A and C, or B and C, showed normal pollen fertility, the accession being classified as type AC or BC. The accessions whose F1 hybrids always showed pollen fertility of less than 75% in all three cross combinations were designated as type X. Kawase et al. (1997) further investigated collections from northern Pakistan. Kawase and Fukunaga (1999) also used a landrace from Lan Hsü Island of Taiwan, which was classified as type X in Kawase and Sakamoto (1987), as tester D, and crossed it with landraces and reclassified type X in Kawase and Sakamoto (1987). Geographical distribution of these different types is shown in Table 7.1. Most type A accessions were distributed in East Asia, including Japan, Korea, and China. Type B accessions were found to a lesser extent in Taiwan and the southwestern part of the Nansei Islands of Japan. Most European accessions were found to be type C. Type D is distributed in Lan Hsü Island of Taiwan and Batan Islands of the Philippines. Accessions of types AC and BC were distributed in Afghanistan and India, respectively. Kawase and Sakamoto (1987) and Sakamoto (1987) concluded that types AC and BC are thought to be less specialized genetically than types A, B, or C, and that the geographical distribution of these landrace groups suggests that *S. italica* was first domesticated in an area ranging from Afghanistan to India, and then dispersed both eastward and westward from there.

In the late 1990s, DNA markers became available for phylogenetic studies. Several studies using DNA markers such as ribosomal DNA (rDNA: Fukunaga et al. 1997, 2005, 2006, 2011; Eda et al. 2013), RFLP (Fukunaga et al. 2002), AFLP (Le Thierry d'Ennequin et al. 2000), TD (Hirano et al. 2011), and SNPs (Jia et al. 2013) have been carried out to clarify the genetic structure of foxtail millet.

7.2.2.1 Ribosomal DNA (rDNA)

Fukunaga et al. (1997) investigated RFLP and the structure of ribosomal RNA genes (rDNA) in 117 landraces of foxtail millet. Five RFLP phenotypes were found when the genomic DNA was digested with *Bam*HI; these were named types I–V. Of these, types I, II, and III were the most frequent. Type I was mainly distributed in the temperate zone, type II in the Taiwan-Philippines Islands, and type III in South Asia. Restriction

Table 7.1 Studies on genetic structure of foxtail millet landraces and geographical groups and center of diversity revealed by the studies

Genetic markers/intraspecific hybrid pollen sterility	Geographical group	Center of diversity	References
Esterase isozymes	East Asia versus Europe	East Asia	Kawase and Sakamoto (1984)
Ten isozymes	China-Korea-Japan, Okinawa (Nansei Islands of Japan)-Taiwan, India-Kenya, Europe		Jusuf and Pernes (1985)
Prolamine	Europe, tropical groups	China	Nakayama et al. (1999)
Hybrid sterility	China-Korea-Japan, Okinawa (Nansei Islands of Japan)-Taiwan, Lan-Hsu Island-Batan Islands India, Afghanistan, Europe		Kawase and Sakamoto (1987), Kawase and Fukunaga (1999)
rDNA	Okinawa (Nansei Islands of Japan)-Taiwan-the Philippines, India, Afghanistan	China	Fukunaga et al. (1997, 2006), Eda et al. (2013)
Nuclear RFLP	East Asia, Nansei Islands-Taiwan-the Philippines, India, Afghanistan-Central Asia-Europe	China	Fukunaga et al. (2002)
mtDNA	Not clear	China	Fukunaga and Kato (2003)
RAPD	Central Europe and two Asiatic groups (north and south)		Schontz and Rether (2000)
AFLP	Not clear	China	Le Thierry d'Ennequin et al. (2000)
TD	East Asia, Nansei Islands-Taiwan-the Philippines, India, Central Asia, Europe	China	Hirano et al. (2011)
SNPs	North China-South China		Jia et al. (2013)

mapping of the cloned rDNA and comparison with RFLP phenotypes showed that the different types originated from a polymorphism in length within the intergenic spacer (IGS) and *Bam*HI site changes within the IGS. Schontz and Rether (1998) also investigated rDNA in a world collection (43 accessions) for variation in repeat unit length and restriction enzyme site variability. They detected two lengths of repeat units of about 7.9 or 7.6 kb; the central European accessions and most western European accessions have only the 7.6-kb repeat unit, and most Asiatic lines have a 7.9-kb repeat unit, although lines originating from the north or the south of Asia showed different numbers of *Bam*HI fragments. These types correspond to types I–III in Fukunaga et al. (1997). They concluded that the fact that the difference between the Asiatic and European pools is not continuous (7.9 or 7.6 kb) excludes the hypothesis of domestication being based on the spread of an initial population over Eurasia.

Fukunaga et al. (1997) suggested that foxtail millet landraces could be differentiated into two main geographical groups, 7.6-kb repeat unit (=type I) from the temperate region and 7.9-kb repeat unit (=types II and III) from the subtropical-tropical region, whereas Schontz and Rether (1998) insisted that foxtail millet landraces differentiate into 7.9-kb repeat unit from Asia and 7.6-kb repeat unit from Europe. The difference between the conclusions of these two studies is because of the difference in the number of Asian accessions used. Fukunaga et al. (2005) also determined the sequence of ribosomal DNA (rDNA) IGS of foxtail millet isolated in the previous study and identified subrepeats in the polymorphic region. Fukunaga et al. (2006) sequenced ribosomal DNA IGS subrepeats and their flanking regions of foxtail millet landraces from various regions in Europe and Asia, as well as its wild ancestor green foxtail, to elucidate phylogenetic differentiation within each of types I–III found in the previous work and to elucidate relationships between these three types. Type I was classified into seven subtypes designated as Ia–Ig based on subrepeat sequences; C repeats downstream of those subrepeats were also polymorphic. Type II was also highly polymorphic, and four subtypes were found and designated as subtypes IIa–IId, but sequence analyses indicated type III as monomorphic. This work indicates that type III should be classified as a subtype of type II (subtype IIe). Sequence polymorphism of subrepeats of types I–III indicated that subrepeats of subtype IIa are very divergent from the others. Relationships between types I–III were much more complicated than anticipated based on previous RFLP work. Recently, the rDNA PCR–RFLP of foxtail millet germplasm (480 accessions) collected throughout Eurasia and from part of Africa was investigated with 5 restriction enzymes (Eda et al. 2013). Foxtail millet germplasm was classified by length of the rDNA IGS and RFLP, and clear geographical differentiation was observed between East Asia, the Nansei Islands of Japan-Taiwan-the Philippines area, South Asia, and Afghanistan-Pakistan (Table 7.1). Evidence of migration of foxtail millet landraces between the areas was also found. Diversity indices (D) for each region were calculated, and it was concluded that the center of diversity of this millet is East Asia, including China, Korea, and Japan.

7.2.2.2 RAPD Markers and AFLP

Schontz and Rether (1999) investigated RAPDs in 37 accessions of cultivated *Setaria italica*, representative of Eurasian accessions. By using four 10-mer primers, they obtained 25 polymorphic bands and identified 33 different genotypes. A factorial analysis of correspondence was performed on the presence-absence data, and three genetic groups were identified. These genetic groups were closely related to the geographical origin of the different accessions: one central European and two Asiatic groups (the first Asiatic accessions originating in latitudes below 35° N and the second comprising the Asiatic accessions originating in latitudes above 35° N) (Table 7.1).

Le Thierry d'Ennequin et al. (2000) investigated AFLP markers to assess genetic diversity and patterns of geographic variation among 39 accessions of foxtail millet and 22 accessions of its wild progenitor. A high level of polymorphism was observed. Dendrograms based on Nei and Li distances from a neighbor-joining procedure were constructed using 160 polymorphic bands. In contrast to other molecular marker studies, no specific geographic structure could be extracted from the data. The high level of diversity among Chinese accessions was consistent with the hypothesis of a center of domestication in China (Table 7.1).

7.2.2.3 Nuclear Genomic and Mitochondrial RFLP

Fukunaga et al. (2002) investigated 16 RFLP loci in 62 landraces to study genetic differentiation in foxtail millet. Among 52 bands, 47 were polymorphic among foxtail millet landraces. A dendrogram based on RFLPs was divided into five major clusters (clusters I–V). Clusters I and II contained mainly accessions from East Asia. Cluster III consisted of accessions from subtropical and tropical regions in Asia such as Nansei Islands of Japan, Taiwan, the Philippines, and India, and cluster IV consisted of some accessions from East Asia, an accession from Nepal, and an accession from Myanmar. Cluster V contained accessions from central and western regions of Eurasia such as Afghanistan, Central Asia, and Europe. Chinese landraces were classified into four clusters. These results indicate that foxtail millet landraces have differentiated genetically between different regions and that Chinese landraces are highly variable (Table 7.1).

Mitochondrial DNA (mtDNA) was characterized by RFLPs in 94 accessions (Fukunaga and Kato 2003). Three RFLP patterns were observed by using rice *atp6* as a probe and were designated as types I–III. Differences between types I and II seem to be attributed to recombination between two *atp6* genes. In East and Southeast Asia and Afghanistan, both types I and II were found, whereas type I was predominant in India, Central Asia, and Europe. In China, type III was also found. Chinese accessions showed higher gene diversity than those from other regions

(Table 7.1). This result supported the previous studies on isozymes and nuclear RFLPs.

7.2.2.4 TD Markers

Hirano et al. (2011) investigated genetic structure by TD using 425 accessions of foxtail millet and 12 of the wild ancestor, green foxtail. They used three recently active transposons (*TSI-1*, *TSI-7*, and *TSI-10*) as genome-wide markers and succeeded in demonstrating the geographical structure for foxtail millet. A neighbor-joining dendrogram based on TD grouped the foxtail millet accessions into eight major clusters, each of which consisted of accessions collected from adjacent geographical areas (Table 7.1). Eleven out of 12 green foxtail accessions were grouped separately from the clusters of foxtail millet. These results indicated strong regional differentiations and a long history of cultivation in each region. They also suggest a monophyletic origin of foxtail millet domestication.

7.2.2.5 SNPs

Recently, a large-scale analysis of whole genome SNP in 916 accessions (mainly from China but also including accessions from other regions such as Japan and Korea, Southeast Asia, South Asia, Central Asia, Europe, Africa, and US) was carried out by Jia et al. (2013). They found that the 916 varieties can be clearly classified into 2 divergent groups (spring-sown from type 1 with 292 varieties and summer-sown from type 2) with 624 varieties, and there was a clear geographical distribution of these 2 groups in the Chinese accessions—the majority of type 1 accessions were from northern China and high-altitude areas of northwest China, whereas most of the type 2 accessions originated from central and southern China, which have warmer climates. As for accessions from other countries, they found that varieties from the same geographical regions tended to belong to the same clades in phylogenetic trees (Table 7.1). They concluded that foxtail millet may have a single origin of domestication but that a deep investigation of the wild ancestor is needed. Their results showing that foxtail millet accessions can be divided into northern and southern groups may correspond with the distribution of rDNA types I and II (Eda et al. 2013) and results of other genetic studies, although most of the materials that Jia et al. (2013) used were from China. Further analysis using more accessions from other countries is also required.

7.3 Perspective

At the present time, foxtail is considered as a model crop for abiotic stress tolerance studies, biofuel traits, C_4 photosynthesis, and nutritional biology (Bennetzen et al. (2012), Muthamilarasan and Prasad 2015; Muthamilarasan et al. 2016). Recent studies in phylogeny and association mapping using next-generation sequencing technology (Jia et al. 2013) have updated relationships in foxtail millet and revealed several candidate genes involved in domestication and diversification of landraces in foxtail millet. Further phylogenetic analyses of foxtail millet and its wild ancestor and further analyses of genes involved in domestication should help to clarify foxtail millet domestication and dispersal.

References

Bennetzen JL, Schmutz J, Wang H, Percifield R, Hawkins J, Pontaroli AC, Estep M, Feng L, Vaughn JN, Grimwood J, Jenkins J, Barry K, Lindquist E, Hellsten U, Deshpande S, Wang X, Wu X, Mitros T, Triplett J, Yang X, Ye CY, Mauro-Herrera M, Wang L, Li P, Sharma M, Sharma R, Ronald PC, Panaud O, Kellogg EA, Brutnell TP, Doust AN, Tuskan GA, Rokhsar D, Devos KM (2012) Reference genome sequence of the model plant *Setaria*. Nature Biotechnol 30:555–561

de Wet JMJ, Oestry-Stidd LL, Cubero JI (1979) Origins and evolution of foxtail millet (*Setaria italica*). J d'Agric et de Bot 26:53–64

Eda M, Izumitani A, Ichitani K, Kawase M, Fukunaga K (2013) Geographical variation of foxtail millet, *Setaria italica* (L.) P. Beauv. based on rDNA PCR–RFLP. Genet Res Crop Evol 60:265–274

Fukunaga K, Kato K (2003) Mitochondrial DNA variation in foxtail millet, *Setaria italica* (L.) P. Beauv. Euphytica 129:7–13

Fukunaga K, Domon E, Kawase M (1997) Ribosomal DNA variation in foxtail millet, *Setaria italica* (L.) P. Beauv. and a survey of variation from Europe and Asia. Theor Appl Genet 97:751–756

Fukunaga K, Wang ZM, Kato K, Kawase M (2002) Geographical variation of nuclear Genome RFLPs and genetic differentiation in foxtail millet, *Setaria italica* (L.) P. Beauv. Genet Res Crop Evol 49:95–101

Fukunaga K, Ichitani K, Taura S, Sato M, Kawase M (2005) Ribosomal DNA intergenic spacer sequence in foxtail millet, *Setaria italica* (L.) P. Beauv. and its characterization and application to typing of foxtail millet landraces. Hereditas 142:38–44

Fukunaga K, Ichiatani K, Kawase M (2006) Phylogenetic analysis of rDNA intergenic spacer subrepeats and its implication for domestication history of foxtail millet, *Setaria italica*. Theor Appl Genet 113:261–269

Harlan JR (1975) Crops and man. Soc Am Agron Crop Sci Soc America, Madison, WI

Hirano R, Naito K, Fukunaga K, Watanabe KN, Ohsawa R, Kawase M (2011) Genetic structure of landraces in foxtail millet (*Setaria italica* (L.) P. Beauv.) revealed with transposon display and interpretation to crop evolution of foxtail millet. Genome 54:498–506

Hunt HV, Linden MV, Liu X, Motuzaite-Matuzeviciute G, Colledge S, Jones MK (2008) Millets across Eurasia: chronology and context of early records of the genera *Panicum* and *Setaria* from archaeological sites in the Old World. Veg Hist Archaeobot 17:5–18

Jia G, Huang X, Zhi H, Zhao Y, Zhao Q, Li W, Chai Y, Yang L, Liu K, Lu H, Zhu C, Lu Y, Zhou C, Fan D, Weng Q, Guo Y, Huang T, Zhang L, Lu T, Feng Q, Hao H, Liu H, Lu P, Zhang N, Li Y, Guo E, Wang S, Wang S, Liu J, Zhang W, Chen G, Zhang B, Li W, Wang Y, Li H, Zhao B, Li J, Diao X, Han B (2013) A haplotype map of genomic variations and genome-wide association studies of agronomic traits in foxtail millet (*Setaria italica*). Nat Genet 45:957–961

Jusuf M, Pernes J (1985) Genetic variability of foxtail millet (*Setaria italica* P. Beauv.). Theor Appl Genet 71:385–393

Kato S, Kosaka H, Hara S (1928) On the affinity of rice varieties shown by the fertility of hybrid plants. Rep Bul Sci Fak Terkult Kyushu Imp Univ. 3:132–147 (in Japanese with English summary)

Kawase M, Fukunaga K (1999) Distribution of Type D, a landrace group newly determined by means of hybrid sterility in foxtail millet, *Setaria italica* (L.) P. BEAUV. Breed Res 1:302 (in Japanese)

Kawase M, Sakamoto S (1984) Variation, geographical distribution and genetical analysis of esterase isozymes in foxtail millet, *Setaria italica* (L.) P. Beauv. Theor Appl Genet 67:529–533

Kawase M, Sakamoto S (1987) Geographical distribution of landrace groups classified by hybrid pollen sterility in foxtail millet, *Setaria italica* (L.) P. Beauv. J. Jpn Breed 37:1–9

Kawase M, Ochiai Y, Fukunaga K (1997) Characterization of foxtail millet, *Setaria italica* (L.) P. Beauv. in Pakistan based on intraspecific hybrid pollen sterility. Breed Sci 47:45–49

Kihara H, Kishimoto E (1942) Bastarde zwischen *Setaria italica* und *S. viridis*. Bot Mag 20:63–67 (in Japanese with German summary)

Le d'Ennequin MLT, Panaud O, Toupance B, Sarr A (2000) Assessment of genetic relationships between *Setaria italica* and its wild relative *S. viridis* using AFLP markers. Theor Appl Genet 100:1061–1066

Li Y, Wu SZ (1996) Traditional maintenance and multiplication of foxtail millet (*Setaria italica* (L.) P. Beauv) landraces in China. Euphytica 87:33–38

Li HW, Li CH, Pao WK (1945) Cytological and genetical studies of the interspecific cross of the cultivated foxtail millet, *Setaria italica* (L.) Beauv., and the green foxtail millet, *S. viridis* L [1945]. J Am Soc Agron 37:32–54

Li Y, Cao YS, Wu SZ, Zhang XZ (1995a) A diversity analysis of foxtail millet (*Setaria italica* (L.) P. Beauv.) landraces of Chinese origin. Genet Resour Crop Evol 45:279–285

Li Y, Wu SZ, Cao YS (1995b) Cluster analysis of an international collection of foxtail millet (*Setaria italica* (L.) P. Beauv). Euphytica 83:79–85

Muthamilarasan M, Prasad M (2015) Advances in *Setaria* genomics for genetic improvement of cereals and bioenergy grasses. Theor Appl Genet 128:1–14

Muthamilarasan M, Dhaka A, Yadav R, Prasad M (2016) Exploration of millet models for developing nutrient rich graminaceous crops. Plant Sci 242:89–97

Nakayama H, Namai H, Okuno K (1999) Geographical variation of the alleles at the two prolamin loci, *Pro1* and *Pro2*, in foxtail millet, *Setaria italica* (L.) P. Beauv Genes Genet Syst 74:293–297

Nasu H, Momohra A, Yasuda Y, He J (2007) The occurrence and identification of *Setaria italica* (L.) P. Beauv. (foxtail millet) grains from the Chengtoushan site (ca. 5800 cal B.P.) in central China, with reference to the domestication centre in Asia. Veget Hist Archaeobot 16:481–494

Sakamoto S (1987) Origin and dispersal of common millet and foxtail millet. Japan Agr Res Quart 21:84–89

Schontz D, Rether B (1998) Genetic variability in foxtail millet, *Setaria italica* (L.) P. Beauv.—RFLP using a heterologous rDNA probe. Plant Breed 117:231–234

Schontz D, Rether B (1999) Genetic variability in foxtail millet, *Setaria italica* (L.) P. Beauv: identification and classification of lines with RAPD markers. Plant Breed 118:190–192

Stevens CJ, Murphy C, Roberts R, Lucas L, Silva F, Fuller DQ (2016) Between China and South Asia: a middle Asian corridor of crop dispersal and agricultural innovation in the Bronze Age. Holcene 1–15

Vavilov NI (1926) Studies on the origin of cultivated plants. Inst Appl Bot Plant Breed 16:1–248

Wang RL, Wendel JF, Dekker JH (1995) Weedy adaptation in Setaria spp. I. Isozyme analysis of genetic diversity and population genetic structure in *Setaria viridis*. Am J Bot 82:308–317

Genetic Determinants of Abiotic Stress Tolerance in Foxtail Millet

8

Charu Lata and Radha Shivhare

Abstract

Foxtail millet is one of the most important C_4 Panicoid crops known for its small genome size (~490 Mb), short life cycle, inbreeding nature, and remarkable abiotic stress tolerance properties. It is a widely-grown food and fodder crop in the dry and semi-arid regions of Asia and Africa, including North China and India. *Setaria italica* (cultivated) and *Setaria viridis* (wild) are two widely known species of *Setaria* genus that serve as excellent model systems for evolutionary, architectural, and physiological studies in related potential bioenergy Panicoid grasses such as switch grass, napier grass, and pearl millet. Foxtail millet is rich in genetic diversity, with several core and mini core collections of its diverse germplasm. There are significant phenotypic variations that provide scope for association mapping and allele mining of new variants of abiotic stress tolerance that could be effectively utilized for crop improvement. Several of the foxtail millet accessions could also be abiotic stress tolerant particularly to drought and salinity, and exploiting their agronomic and stress tolerant traits could be particularly important for marker-assisted selection and genetic engineering. Furthermore, with the release and availability of the foxtail millet genome sequence, several of its distinctive attributes, including abiotic stress tolerance, have been discovered that may help in a better understanding of its evolution, stress physiology, and adaptation. The foxtail millet genome sequence thus not only helps toward identification and introgression of agronomically important traits but also helps in deciphering the abiotic stress tolerance mechanisms of this exceptionally stress tolerant crop and is also useful in developing climate resilient crops which are very crucial in this era of global climate change.

Charu Lata (✉) · Radha Shivhare
CSIR-National Botanical Research Institute,
Rana Pratap Marg, Lucknow 226001, India
e-mail: charulata@nbri.res.in;
charulata14@gmail.com

Charu Lata · Radha Shivhare
Academy of Scientific and Innovative Research
(AcSIR), Anusandhan Bhawan, 2 Rafi Marg,
New Delhi 110 001, India

M. Prasad (ed.), *The Foxtail Millet Genome*, Compendium of Plant Genomes,
DOI 10.1007/978-3-319-65617-5_8

8.1 Introduction

Millets that represent small-grained cereal crops are among the developing world's major crops grown for human consumption and livestock feed, including pearl millet, foxtail millet, finger millet, proso millet, little millet, kodo millet, Japanese and Indian barnyard millet, teff, and fonio. They have typically been originated, domesticated, and cultivated in semi-arid and tropical regions of Africa and Asia (Dvořáková et al. 2015). However, they are also grown in parts of America and Eurasia (Goron and Raizada 2015). Millets also play a crucial role in the economy of developing countries where a large acreage of marginal land is used for cultivation. Notably, India contributes significantly to global millet production with around 30% of total world millet production obtained from 25% of the total world area under millet production in 2013, indicating the importance of millets for the resource poor and marginal farmers of the Indian subcontinent (Tadele 2016). Millets distinctively are well-adapted to adverse agro-ecological conditions, are typically grown in marginal soils with the least amounts of input yet still rich in nutrients, and are a source of staple dietary supplements for millions of inhabitants of these regions. Interestingly, millets exhibit sufficient variability among themselves, which is not only exhibited in terms of morphology such as plant stature, seed color and size, panicle size, etc. but also demonstrated at the genetic level in terms of chromosome number and ploidy level that ranges from $2n = 2x = 14$ in pearl millet to $2n = 6x = 54$ in fonio (Tadele 2016). They have been broadly grouped into two major subfamilies, namely Panicoideae that comprises pearl millet, foxtail millet, proso millet, little millet, and Japanese barnyard millet and Chloridoideae that includes finger millet and teff, belonging to eight genera, namely *Pennisetum, Setaria, Echinocloa, Elusine, Panicum, Paspalum, Eragrostis,* and *Digitaria.* Thus, agronomically the Panicoideae subfamily is the most important grass family as it includes not only the most economically important C_4 cereal crop maize but also sorghum and most of the small-grained millets. Sugarcane, a major biofuel crop and *Miscanthus* and *Panicum virgatum* (switchgrass), the emerging bioenergy feedstocks, also belong to this subfamily (Li and Brutnell 2011). Both *Setaria italica* (foxtail millet) and *Setaria viridis* (green foxtail) are the closest relatives of the important bioenergy feedstock, switch grass, and thus are considered excellent models to study the evolution, architecture, and physiology of switchgrass and related C_4 Panicoid grasses (Lata et al. 2013).

Foxtail millet is a self-pollinating, diploid ($2n = 2x = 18$), C_4 Panicoid small-grained millet crop with a very small genome size ($\sim$515 Mb) and short life cycle (Lata et al. 2013). It is one of the world's oldest cultivated grain crops, domesticated $\sim$8,000 years ago in Northern China (Li and Wu 1996; Lata et al. 2013). It ranks second in total global millet production and is presently one of the minor food crops of dry regions of Southern Europe and Asia, including India and China, whereas in North America it is primarily grown for silage, bird feed, and fodder or cover crop (Goron and Raizada 2015). Foxtail millet grains are nutritionally very rich, with comparatively higher seed protein, crude fat, iron, and mineral content as compared to staple cereal crops such as rice and wheat (Zhang et al. 2007a; Lata et al. 2013). Its grains are also high in fiber content (Amadou et al. 2011) and its bran is enriched with linoleic and oleic acids (Liang et al. 2010). Additionally, foxtail millet is widely known for its exceptional drought tolerance and high water use efficiency (WUE) (Li and Brutnell 2011; Lata et al. 2011a), and some of its cultivars are also found to be salt tolerant (Jayaraman et al. 2008). Considering both its agronomic and economic importance, several research groups across the world have recently focused their attention on the development of genetic and genomic resources for this crop as well as on understanding the physiological and molecular basis of its excellent abiotic stress tolerance characteristics. Much work has now been done on exploring existing germplasm resources for identification of newer alleles and stress resistance traits, as well as generation and utilization of genomic resources for introgression of agronomically important traits and marker-assisted selection (MAS) for crop improvement and cultivar development. This chapter thus focuses on the recent progress toward unravelling the genetic determinants of abiotic stress tolerance in foxtail millet and their exploitation in crop improvement programs for developing varieties.

8.2 Phylogeny and Genomic Relationship Among *Setaria* Species

Setaria is a diverse genus with approximately 125 species reported to date with different phenotypic traits, life cycles, and ploidy levels (Lata et al. 2013). *Setaria* species have been classified into three gene pools based on genome organization. For example, *S. italica* (domesticated and cultivated) and *S. viridis* (wild ancestor) which are considered very important from genetic studies perspective and represent the primary gene pool have AA genome (Benabdelmouna et al. 2001). *S. italica* was domesticated from *S. viridis* around 8,000 years ago in Northern China (Barton et al. 2009). *S. faberi* (giant foxtail or nodding millet) and *S. verticillata* (bristly foxtail) with AABB genome form the secondary gene pool and are thought to originate from a cross between *S. viridis* (AA) and *S. adhaerens* (bristly grass; BB) (Benabdelmouna et al. 2001). The remaining *Setaria* landraces such as *grisebachii, queenslandica, pumila,* etc. constitute the tertiary gene pool. Several phylogenetic and cytogenetic studies suggested that *S. viridis, S. adhaerens, S. faberi, S. verticillata,* and *S. pumila* are discrete taxa (Benabdelmouna et al. 2001; Benabdelmouna and Darmency 2003; Layton and Kellogg 2014). *S. pumila*, whose genome organization is not yet known, is confirmed to be unrelated despite its widespread habitat similarity with other *Setaria* species (Layton and Kellogg 2014). The *Setaria* genus complexity has been reviewed in detail by Lata et al. (2013). A detailed understanding of the complex and intricate phylogenetic and genomic relationships among *Setaria* species would be beneficial for identification and selection of germplasm with desired traits, as well as for competent parent selection and hybridization required for improved breeding strategies.

8.3 Population Structure and Genetic Diversity Among Foxtail Millet Accessions

S. viridis has been projected as a Panicoid grass model because of its distinctive attributes such as small stature, diploidy, self-pollination, short generation time, minimal growth requirements, and excellent genetic transformation system that make it an exceptional model for genetic and evolutionary analyses. It also closely shares several of its cell wall characteristics with Panicoid grasses as compared to rice (Brutnell et al. 2015). Efforts have therefore recently been made to assess the genetic diversity of both wild and domesticated forms of *Setaria* in natural populations (Wang et al. 2010; Huang et al. 2014; Gupta et al. 2014). Various investigations on *Setaria* population structure showed the existence of subpopulations associated with distinct geographical locations (Brutnell et al. 2015). For example, an earlier population genetic study of 168 *S. viridis* and *S. italica* accessions from Eurasia and North America could not differentiate between the accessions from these two geographical locations but could distinctly identify southern and northern populations from central North America (Wang et al. 1995). Another genetic diversity study of a worldwide collection of 200 *S. viridis* accessions using genotyping-by-sequencing) and *S. italica* genome as reference revealed two distinct groups of *S. viridis* and a third group comprising the Chinese *S. viridis* accessions that resembled *S. italica*-like accessions, and all three groups also showed considerable admixture among themselves owing to rapid linkage disequilibrium (LD) decay in the overall sample (Huang et al. 2014). Intriguingly, the study reported a very strong correlation between genetic distance and climate, and genetic distance and geography in the North American *S. viridis* accessions, suggesting a balance between genetic drift and gene flow and multiple introductions and/or local adaptation to climate. A rapid decay in LD within 150 bp in wild green foxtail was also reported earlier (Wang et al. 2010). The authors reported that the level of LD in domesticated foxtail millet extends up to 1 kb. The genetic diversity study was carried out across 9 loci in 50 domesticated and 34 wild foxtail millet accessions that suggested a low level of genetic diversity in wild foxtail and also that domestication bottleneck of foxtail millet is more severe compared to maize and somewhat less prominent compared to rice (Wang et al. 2010). Another population structure and relative kinship study involving 184 foxtail millet accessions from diverse Indian geographical locations and 50 simple sequence repeat (SSR) markers showed a significant association (R^2 = 18%) between 8 SSR markers and 9 agronomic traits, including flag leaf width and grain yield (Gupta et al. 2014).

8.4 Identification of Germplasm with Stress Resistance Traits

8.4.1 Biotic Stress Resistance

Millet crops are usually infected by fungal diseases and to a lesser extent by bacterial and viral diseases (Dwivedi et al. 2012). Blast, downy mildew, rust, and smut are the common fungal diseases that infect foxtail millet. Recently a new leaf and sheath brown spot fungus *Bipolaris australiensis* has also been reported to infect this crop (Mirzaee et al. 2010). Bacterial blight caused by *Xanthomonas* spp. is the most important bacterial disease affecting foxtail millet. This crop is also a carrier for both wheat curl mite *(Eriophyes tullipae* Keifer), which is a carrier of wheat streak mosaic virus and the virus itself (Baltensperger 2002). This viral disease, however, does not impact the already adapted foxtail millet cultivars, but they act as an over-summering host and severely affect nearby wheat fields. The effects of the above-mentioned diseases range from mild symptoms to severe infections, leading to devastations when large swathes of land are affected. However, not much work has been done regarding disease management in foxtail millet compared to pearl millet.

There has been a recent report where 17 foxtail millet germplasms were screened for leaf blast and rust at Regional Agricultural Research Station, Andhra Pradesh, India (Munirathnam et al. 2015). The study suggested DHFtMV 2-5 to be moderately resistant to both leaf blast and rust and it could be recommended for blast and rust prone areas. In an earlier study a core collection of 155 foxtail millet accessions were screened for new and diverse sources of blast resistance against Patancheru isolate (Fx 57) of *Magnaporthe grisea* (Sharma et al. 2014). The study reported two accessions, namely ISe 1181 and ISe 1547, to be free from head blast infection as well as resistance to neck, leaf, and sheath blast. On the other hand, ISe 1067 and ISe 1575 showed high levels of resistance to blast. However, in an earlier study, preliminary genetic analysis of four Japanese fungus isolates from foxtail millet suggested that blast resistance may be controlled by more than two dominant genes in this crop (Nakayama et al. 2005). Several foxtail millet germplasms have been identified and advanced lines for wheat streak mosaic resistance with enlarged head size and improved yield have also been developed (Siles et al. 2001).

8.4.2 Abiotic Stress Tolerance

Although foxtail millet is generally well-adapted to various abiotic stresses including drought, salinity, high temperature, and poor soil, even then this crop is not totally immune to abiotic stresses. Several of its varieties and accessions are also prone to various abiotic stresses. For example, foxtail millet being thinner stemmed is badly affected by lodging (Dwivedi et al. 2012). The worldwide foxtail millet accessions preserved in national and international genebanks can serve as an excellent resource for identification and exploitation of genetic variations for abiotic stress tolerance through precise phenotyping that can ultimately be utilized for crop improvement programs aimed at developing stress tolerant crops. In recent years foxtail millet has garnered substantial attention from the research community because of its projection as an excellent model for studying architecture, evolution, physiology, and stress tolerance attributes (Lata et al. 2013). Considerable work has been done toward germplasm identification and understanding the genetic basis of abiotic stress tolerance, particularly drought, salinity, lodging, and water logging in foxtail millet. Some progress has been made toward understanding the physiology of abiotic stress tolerance and the genomic regions associated with drought stress tolerance.

8.4.2.1 Drought

Foxtail millet is known to be an abiotic stress tolerant crop, particularly to drought, and its WUE is reported to be higher than that of maize, wheat, and sorghum (Shantz and Piemeisel 1927; He and Bonjean 2010). Its high WUE and short life cycle make it an elite drought-tolerant crop (Zhang et al. 2007b; Lata et al. 2011a). However, it is most sensitive to drought at the inflorescence and spikelet development stage (about 35–50 days after sowing). About 17,799 drought-tolerant accessions (17,313 landraces and 486 elite cultivars) of foxtail millet have been divided into five major classes on the basis of their survival capacity in different intensities of drought stress (Li 1991, 1997). In China, researchers had developed screening methods for assessing drought tolerance capacity of foxtail millet germplasms using polyethylene glycol (PEG 6000) or mannitol and identified relative water content and germination rate parameters as indicators of drought stress tolerance (Wen et al. 2005). Foxtail millet genotypes have been screened at the seedling stage for drought tolerance using 20% PEG (Zhang et al. 2005; Zhu et al. 2008). Lata et al. (2011b) screened about 107 foxtail millet cultivars from different geographical locations for drought tolerance at the seedling stage using 20% PEG, and two highly tolerant (cv. IC-403579 and cv. Prasad) and two highly sensitive (cv. IC-480117 and cv. Lepakshi) foxtail millet cultivars were identified. The study also indicated lipid peroxidation as an important marker for drought stress tolerance at the seedling stage in foxtail millet. The identified dehydration-tolerant genotypes may be used for crop improvement purposes (Lata et al. 2011b). Many drought-inducible genes with various functions have been identified on the basis of genomic and molecular functions in foxtail millet (Li et al. 2014; Lata et al. 2010, 2011a, 2014; Muthamilarasan et al. 2014a, b).

8.4.2.2 Salinity

A limited number of reports have been available on soil salinity response in foxtail millet, unlike other cereals. The shoot Na^+ concentration could be considered as a potential nondestructive selection criterion for vegetative-stage screening (Krishnamurthy et al. 2007). Zhi et al. (2004) has screened 260 foxtail millet landraces and cultivars using relative germination rate at 1.0 and 1.5% NaCl concentration and the results showed a high range of variations from 0 to 90% in different cultivars and landraces (0–20% in 29 accessions, 21–50% in 45 accessions, 51–90% in 153 accessions, and over 90% in 33 accessions). On salinity stress, genes of glutamine synthetase (GS) and pyrroline-5-carboxylate (P5C) reductase were found to be up-regulated, resulting in higher transcript levels of these two enzymes, which play an important role in the biosynthesis of proline, a molecule for osmotic regulation in stressed plants (Huang et al. 2013). Sreenivasulu et al. (1999) checked the peroxidase activity on two cultivars of foxtail millet and categorized them as salt tolerant (cv. Prasad) and salt sensitive (cv. Lepakshi). Furthermore, Sreenivasulu et al. (2004) identified 620 differentially expressed-sequence-tags (ESTs) from cv. Prasad and cv. Lepakshi, representing unigenes of a barley EST collection using a 711-cDNA inserts macroarray filter. Among these transcripts, hydrogen peroxide-scavenging enzymes including phospholipid hydroperoxide glutathione peroxidase (PHGPX), ascorbate peroxidase (APX), and catalase 1 (CAT1) as well as different oxidoreductase enzymes such as glutamine synthetase and pyrroline-5-carboxylate (P5C) and some genes of cellular metabolism were found to be significantly up-regulated under high salinity stress in the tolerant cultivar compared to the sensitive. The result showed significantly higher proline content in seedlings of both the cultivars, although the concentration of proline was found to be more in the tolerant than in the sensitive cultivar and also showed a positive correlation with increased glutamine synthetase and P5C reductase activities (Veeranagamallaiah et al. 2007).

To get a better perspective of the salinity stress responses at the molecular level, the temporal changes in total protein profile of cv. Prasad seedlings were examined under different salt stress conditions (Veeranagamallaiah et al. 2008) which led to the identification of 29 differentially-expressed salt-responsive proteins (both up- and down-regulated). Puranik et al. (2011a) analyzed the biochemical responses of 21-day-old seedlings of foxtail millet at 250 mM NaCl stress for 1–48 h for a better understanding of salt stress response in these two cultivars. The two cultivars showed differential salt stress responses when analyzed for lipid peroxidation, and different reactive oxygen scavenging enzymes such as glutathione reductase and catalase under short-term salinity stress. To understand the molecular mechanisms in response to short-term salinity stress, researchers constructed two suppression subtractive hybridization cDNA libraries (forward and reverse) leading to the identification of a total of 249 non-redundant ESTs which were grouped into 11 different categories. cDNA-macroarray analysis of these clones revealed 159 to be differentially expressed under salinity treatment with 115 up- and 44 down-regulated. These transcripts were reported for the first time representing untapped gene sources allowing specific responses to short-term salt-stress in an orphan crop known to possess a natural adaptation capacity to abiotic stress

8.4.2.3 Other Abiotic Stresses

Foxtail millet can also be tolerant to other abiotic stresses such as low temperature, lodging, and water logging. The northern limit of foxtail millet cultivation in China was 50°N. However, researchers in China have developed an extremely-cold-tolerant foxtail millet cultivar Liggu No 26, which helped in extending its cultivation up to 385 km farther north to 54°N (Chen and Qi 1993; Dwivedi et al. 2012). Foxtail millet production is also reportedly being constrained by lodging, resulting in substantial yield losses as well as poor grain quality. According to Tian et al. (2010), a lodging coefficient based on stem and root traits of foxtail millet could be a fitting indicator for estimating lodging resistance in field conditions. The mechanical strength of the stem, heights and weights of the above and underground plant tissues were recommended as the major donors of lodging coefficient across foxtail millet germplasm (Tian et al. 2010; Dwivedi et al. 2012). In another study, path analyses pointed out that breaking strength of stem, associated with greater culm diameter and culm wall thickness, is the most crucial factor determining lodging coefficient (Tian et al. 2015). Longgu 28 and Nenxian 13 are two lodging resistant cultivars that have been developed in China (Dwivedi et al. 2012). Foxtail millet production is also affected by water logging. A waterlogging tolerant foxtail millet cv. Lugu No.7 has been reportedly developed to combat this constraint (Chen and Qi 1993). Considering the above facts, it is apparent that discovery and utilization of untapped novel variations and precise phenotyping for abiotic stress tolerance are key to improving adaptation to adverse environmental conditions in foxtail millet. Furthermore, an improved understanding of the physiological and molecular mechanism(s) of abiotic stress tolerance in foxtail millet can be effectively utilized for developing more stress-tolerant cultivars.

8.5 Factors Responsible for Abiotic Stress Tolerance in Foxtail Millet

8.5.1 Transcription Factors

Plant responses under abiotic stress conditions are a complex phenomenon that has various dynamic responses manifested at physiological, biochemical and molecular levels. Plants subjected to stress environments perceive stress signals through specific receptors and communicate via sophisticated signal transduction pathways, resulting in activation of stress-inducible transcription factors (TFs) and downstream stress-responsive gene expression. TFs associated with abiotic stress responses in foxtail millet are listed in Table 8.1. To open the insights of dehydration stress regulatory pathway in foxtail millet, a comparative transcriptome analysis has been carried in a drought-tolerant cultivar cv. Prasad using a suppression subtractive hybridization (SSH) technique. This study resulted in identification of a novel stress- responsive TF *SiDREB2* belonging to DREB (dehydration-responsive element-binding proteins) sub-family (Lata et al. 2010). The transcript level of *SiDREB2* was found to be significantly up-regulated in tolerant cultivar, suggesting this gene might play an important role in stress-responsive mechanisms in foxtail millet. DREB is an important subfamily of AP2/ERF (APETALA2/ethylene-responsive element-binding factor) TFs and participates in the regulation of stress-responsive gene expression through abscisic acid (ABA)-independent pathways (Lata and Prasad 2011).

A novel *SiDREB2* has been cloned and characterized from foxtail millet and used to develop an allele-specific marker (ASM) for dehydration tolerance (Lata et al. 2011a; Lata and Prasad 2013b, 2014). Furthermore, to understand the role of AP2/ERF TFs in foxtail millet, a genome-wide analysis has been carried out using in silico approaches (Lata et al. 2014). A total of 171 AP2/ERF encoding genes in the *S. italica* genome have been identified, of which 48 were DREB TFs evaluated with the help of phylogenetic and domain architecture analysis. NAC TFs are also well-known for their regulatory role in biotic as well as abiotic stress in many crop plants, and a subtractive hybridization study in *S. italica* showed significant up-regulation of *SiNAC* in salinity stress libraries (Puranik et al. 2011a). Puranik et al. (2011b) reported the cloning of *SiNAC*, a novel membrane-associated NAC gene that functions as an activator of transcription in stress responses and also regulates plant developmental pathways. Puranik et al. (2013) also carried out genome-wide identification and characterization of NAC TFs in *S. italica*. Among the

Table 8.1 Details of important transcription factors linked with abiotic stress responses in foxtail millet

Transcription factor	Numbers identified	Stress response	Functional characterization	References
SiDREB2	1	Dehydration, drought and salt response	Yes	Lata et al. (2011a)
SiNAC2	1	Dehydration, drought and salt response	Yes	Puranik et al. (2011b)
SiNAC	147	Drought and salt stress	–	Puranik et al. (2013)
SiAP2/ERF	171	Drought and salinity response	–	Lata et al. (2014)
SiARDP	1	Drought	Yes	Li et al. (2014)
SiMYB	209	Abiotic stress and hormone response	–	Muthamilarasan et al. (2014b)
SiC2H2	124	Abiotic and hormone response		Muthamilarasan et al. (2014a)
SiWRKY	105	Dehydration, salt and hormone stress	–	Muthamilarasan et al. (2015)
SvWRKY	44	Dehydration, salt and hormone stress	–	Muthamilarasan et al. (2015)
SiNF-YA1 and *SiNF-YB8*	01	Drought and salt stress	Yes	Feng et al. (2015)
SiASR1	01	Drought and oxidative stress	Yes	Feng et al. (2016)
SiDof	35	Drought stress	–	Zhang et al. (2017)

identified 147 *SiNAC* genes, 50 candidate genes were selected for quantitative expression analysis under various abiotic stress treatments. Results of the above study suggested *SiNAC128* as a potential candidate gene for further in-depth characterization (Puranik et al. 2013).

Zhang et al. (2012) and Bennetzen et al. (2012) released a draft genome sequence of *S. italica* that revealed various novel aspects of foxtail genome and facilitated the identification and characterization of a few important stress-responsive TFs, including MYB and C2H2, and miRNAs (Lata and Prasad 2013a). These two (MYB and C2H2 proteins) TFs constitute the largest TF families in plants, playing crucial roles in various developmental and stress-responsive processes (Ambawat et al. 2013). To analyze their role in abiotic stress response, a comprehensive genome-wide study has been performed on foxtail millet (Muthamilarasan et al. 2014a, b). A total of 209 and 124 gene family members of MYB and C2H2, respectively, were identified. Furthermore, on the basis of phylogenetic analysis, *SiMYB* proteins grouped into ten groups (I–X) and *SiC2H2* proteins into five groups (I–V). It has also be found that *SiMYB* and *SiC2H2* protein sequences show significant similarity with their orthologs in sorghum, maize, and rice that show conservation in the overall protein structure of these TFs (Muthamilarasan et al. 2014a, b). Expression patterns of *SiMYB* and *SiC2H2* candidate genes in response to abiotic stresses and hormone treatments using qRT-PCR revealed specific and/or overlapping expression patterns of these genes. Out of analyzed expression profiles of 11 candidate *SiMYB* genes, 3 (*SiMYB124, SiMYB126*, and *SiMYB150*) showed significant up-regulation during drought stress.

In the case of *SiC2H2*, nine candidate genes were selected for expression analysis under abiotic stress response that suggested late expression of *SiC2H2 031*, whereas *SiC2H2 78, SiC2H2 85*, and *SiC2H2 94* showed higher expression during the early phase of drought stress. WRKY proteins also play a significant role in signaling pathways associated with different stress responses. A genome-wide analysis of the WRKY TF family in *S. italica* (*SiWRKY*) and *S. viridis* (*SvWRKY*) was conducted, leading to the identification of 105 *SiWRKY* and 44 *SvWRKY* proteins, respectively (Muthamilarasan et al. 2015). Sequence alignment of these WRKY proteins classified them into three major groups, namely groups I, II, and III. Most of the WRKY proteins fall into group II (53 *SiWRKY* and 23 *SvWRKY*), followed by group III (39 *SiWRKY* and 11 *SvWRKY*) and group I (10 *SiWRKY* and 6 *SvWRKY*). Phylogeny analysis further divided group II into five sub-groups (IIa–e). Comparative mapping of *SiWRKY* and *SvWRKY* genes among related C_4 Panicoid genomes demonstrated the orthologous relationships between these genomes. Expression profiling of candidate *SiWRKY* genes in response to stress (dehydration and salinity) and hormone treatments (ABA, salicylic acid, and methyl jasmonate) suggested the putative involvement of *SiWRKY 066* and *SiWRKY 082* in stress and hormone signaling (Muthamilarasan et al. 2015).

In another study, six ABA stress ripening (ASR) genes were identified from foxtail millet (Feng et al. 2016). The proteins contain ABA/WDS domain and are a class of plant-specific TFs. Overexpression of *SiASR1* in tobacco led to enhanced drought and oxidative stress tolerance as well as altered expression levels of *NtSOD, NtAPX, NtCAT, NtRbohA*, and *NtRbohB* genes, suggesting its important role in stress-related signaling. Furthermore, the expression level of *SiASR4*, a target gene of *SiARDP*, increased under drought and salt stress in transgenic *Arabidopsis* and foxtail millet plants (Li et al. 2017). The *SiASR4* transgenic plants showed enhanced transcription of stress-responsive and reactive oxygen species (ROS) scavenger-associated genes. Together, these findings suggested that *SiASR4* functions in drought and salt stress environments and is regulated by *SiARDP* via an ABA-dependent pathway (Li et al. 2017). Similarly, *SiARDP* an ABA-responsive DRE-binding protein-coding gene enhances drought and salt stress tolerance in Arabidopsis and improves drought stress tolerance in transgenic foxtail millet (Li et al. 2014). The gene was also found to be regulated by two ABA-responsive element binding (AREB)-type TFs, namely *SiAREB1* and *SiAREB2*, which were able to bind SiARDP both in vitro and in vivo physically. Furthermore, in view of the role of TFs in modulating stress-responsive gene regulatory networks, an in silico study was undertaken to identify and characterize total TF-encoding genes in *S. italica* genome (Bonthala et al. 2014). The study identified 2,297 putative TFs and categorized them in 55 families. This information is available in the Foxtail millet Transcription Factor Database (http://59.163.192.91/FmTFDb/) in which complete details of the TFs are compiled, including their sequences, physical positions, tissue-specific gene expression data, gene ontologies, and phylogeny (Bonthala et al. 2014). This database is useful in pinpointing candidate TFs for stress-related studies and performing large-scale investigations.

8.5.2 Other Stress Responsive Genes

Other than TFs, several stress-responsive genes have also been analyzed for their expression profiles and activities under various abiotic stresses in foxtail millet (Table 8.2). Isolation and functional characterization of osmotic stress-responsive genes from foxtail millet could be an important measure for deciphering abiotic stress tolerance mechanisms. Sreenivasulu et al. (2004) examined the expression profile of a phospholipid hydroperoxide

Table 8.2 Functional validation of important genes linked with abiotic stress response in foxtail millet

Gene	Gene function	Source	Plant/organism tested	Type of tolerance	References
SiOPR1	12-Oxophytodienoic acid reductase 1	Foxtail millet	Foxtail millet	Drought tolerance	Zhang et al. (2007b)
Aldose reductase	Detoxify free toxic aldehydes	Foxtail millet	Foxtail millet	Salt tolerance	Veeranagamallaiah et al. (2009)
DNAj	Chaperon protein	Foxtail millet	Wheat	Drought and heat tolerance	Wang et al. (2009)
SiWD40	WD protein	Foxtail millet	Foxtail millet	Salt and Drought tolerance	Mishra et al. (2012)
SiPLDa1	Phospholipase D	Foxtail millet	*Arabidopsis Thaliana*	Drought stress	Peng et al. (2010)
SiARDP	ABA-responsive DRE-binding protein gene	Foxtail millet	*Arabidopsis Thaliana*, Foxtail millet	Salt and Drought tolerance	Li et al. (2014)
SiLEA14	LEA protein	Foxtail millet	*E. coli*, *Arabidopsis Thaliana*, Foxtail millet	Salt and Drought tolerance	Wang et al. (2014a)
SiREM6	Encode remorin protein	Foxtail millet	*Arabidopsis Thaliana*	Salt tolerance	Yue et al. (2014)
SiALDH	Aldehyde dehydrogenase	Foxtail millet	*Escherichia coli*	Salt tolerance	Chen et al. (2014)
SiLTP	Lipid transfer protein	Foxtail millet	Tobacco, Foxtail millet	ABA, salinity and drought stress tolerance	Pan et al. (2016)
SisHSP-27	Heat shock proteins	Foxtail millet	Yeast	Drought, heat, salt and cold	
SiASR4	Encode abscisic acid, stress-and ripening induced proteins	Foxtail millet	*Arabidopsis Thaliana*, Foxtail millet	Salinity and drought tolerance	Li et al. (2017)

glutathione peroxidase gene *PHGPX* in the tolerant foxtail millet cultivar Prasad and proposed its crucial role in conferring salt stress tolerance and in oxidative stress-induced defense reactions. A 12-oxophytodienoic acid reductase 1 gene *SiOPR1* was reported to be highly expressed in foxtail millet roots under drought stress. However, the gene was not found to be influenced by NaCl, ABA, and methyl jasmonate treatments, indicating its important role in drought stress tolerance (Zhang et al. 2007c). The comparative expression profiles of glutamine synthetase (*GS*) and pyrroline-5-carboxylate (*P5C*) reductase under salinity stress in a salt tolerant (cv. Prasad) and a salt sensitive (cv. Lepakshi) foxtail millet cultivar indicated a positive correlation with higher proline accumulation (Veeranagamallaiah et al. 2007). An improved aldose reductase activity with increasing salt stress was also reported in these two foxtail millet cultivars (Veeranagamallaiah et al. 2009). Interestingly, the increase in the enzyme activity was positively correlated with accumulation of sorbitol, essential for osmotic balance, and 4-hydroxynon-2-enal, a major product of lipid peroxidation.

Zhao et al. (2009) reported improved aluminum stress tolerance in transgenic Arabidopsis plants overexpressing *Si69*, a Wali7 homologue. In another study, foxtail millet *DNAj* gene, associated with drought and heat tolerance, was introgressed into four wheat cultivars via the pollen-tube pathway, paving the way for developing drought-tolerant wheat lines (Wang et al. 2009). Overexpression of a phospholipase D gene *SiPLDα1* in Arabidopsis remarkably enhanced drought tolerance of transgenic plants (Peng et al. 2010). The Arabidopsis transgenic plants had higher biomass, increased relative water content, reduced electrolytic leakage, and higher survival percentages, with no undesirable effect on their growth and development compared to control plants. Augmented expression levels of several stress-related genes were also observed in the *35S::SiPLDα1* transgenic Arabidopsis plants, suggesting *SiPLDα1* as a useful target gene for improving drought stress tolerance (Peng et al. 2010). Mishra et al. (2012) reported a nuclear-localized WD-repeat-containing protein gene *SiWD40* to be induced under various abiotic stresses such as dehydration, salt, cold, and ABA. It was also suggested that an AP2-domain-containing protein SiAP2 might regulate the environmental stress-responsive *SiWD40* expression. Recently, 11 plant-specific remorin protein family genes were identified from foxtail millet (Yue et al. 2014). The remorin proteins have the ability to attach to the plasma membrane (Jacinto et al. 1993). One of the remorin family genes, *SiREM6*, was found to improve high salt stress in transgenic Arabidopsis at germination and early seedling growth stages. Interestingly, the *SiREM6* promoter contained two DRE elements and one AREB element. However, only a DRE-binding TF *SiARDP* could physically bind to the DRE elements of the *SiREM6* promoter whereas *SiAREB1* could not bind to the AREB-element, indicating *SiREM6* to be a target gene of *SiARDP* (Yue et al. 2014).

In a similar kind of study, researchers have characterized a late embryogenesis abundant (LEA) gene *SiLEA14* from foxtail millet (Wang et al. 2014a). The LEA proteins are known to play important roles in protecting plants from various abiotic and biotic stresses. *SiLEA14* gene was expressed in different developmental stages including roots, stems, leaves, inflorescence, and seeds. The gene was found to be highly induced by salt stress and exogenous ABA application. Overexpression of *SiLEA14* improved *Escherichia coli* growth performance under salt stress as compared to control. The Arabidopsis seedlings overexpressing *SiLEA14* exhibited enhanced salt and osmotic stress tolerance in comparison to the wild-type (WT) plants. The transgenic foxtail millet seedlings also displayed better growth under salt and drought stresses than the WT. The results thus indicated *SiLEA14* to be a novel atypical LEA protein that might play significant roles in abiotic stress resistance of crop plants. One of the studies highlighted the role of aldehyde dehydrogenases (ALDH) in various abiotic stress responses such as osmotic stress, cold, H_2O_2, and ABA in foxtail millet. ALDH genes are known to detoxify ROS indirectly and help in reducing lipid peroxidation mediated cellular toxicity under various environmental stresses (Chen et al. 2014). The genome-wide survey led to the identification of 20 *SiALDH* genes. Expression analysis suggested organ- and stress-specific expression of these genes in foxtail millet, and the transformation of *SiALDH2B2*, *SiALDH10A2*, *SiALDH5F1*, *SiALDH22A1*, and *SiALDH3E2* genes in *E. coli* led to improved tolerance to salt stress (Chen et al. 2014). A similar investigation for identification and characterization of cytokinin oxidase/dehydrogenase (CKX) from foxtail millet was performed (Wang et al. 2014b). The results indicated high up-regulation of *SiCKX* genes under various abiotic stresses, hinting at their possible role in stress regulation.

Pan et al. (2016) isolated a non-specific lipid transfer protein (LTP) coding gene *SiLTP* from foxtail millet. The LTPs are low molecular weight cysteine-rich soluble proteins with diverse roles in different developmental stages as well as in biotic and abiotic stress responses. The *SiLTP*-overexpressing foxtail millet lines showed improved tolerance to salt and drought stresses whereas the *SiLTP* RNA interference (RNAi)-based transgenic lines of foxtail millet were found to be more sensitive to salt and drought stress compared to control. Recently, genome-wide investigation for various heat shock proteins (HSPs) have also been performed in foxtail millet, leading to the identification of 20, 9, 27, 20, and 37 genes belonging to *SiHSP100*, *SiHSP90*, *SiHSP70*, *SiHSP60*, and *SisHSP* families, respectively (Singh et al. 2016). Expression profiling indicated up-regulation of several HSPs in the tolerant cultivar compared to the sensitive one under dehydration, heat, salinity, and cold stresses. Furthermore, overexpression of a small HSP gene *SisHSP-27* in yeast conferred tolerance to various abiotic stresses.

Altogether, these studies have identified potential candidate genes that could be effectively utilized in crop improvement programs for improving abiotic stress tolerance. However, the lack of an efficient transformation system for expressing the candidate genes in foxtail millet remains a bottleneck in *Setaria* genomics, even though these genes are reported to be effective in enhancing stress tolerance of transgenic plants by regulating the expression of broad-spectrum stress-related genes. As a consequence, the detailed molecular, cellular, and physiological mechanisms responsible for variation in tolerance to various abiotic stresses, including drought tolerance among foxtail millet lines, have not yet been fully elucidated.

8.5.3 Small RNAs

Small RNAs (sRNAs) are noncoding RNAs, including miRNA, siRNA, snRNA, snoRNA, etc. (Sunkar and Zhu 2004). Among them, microRNAs (miRNAs) and endogenous small interfering RNAs (siRNAs) represent two major classes of small RNAs involved in gene regulation. Both miRNAs and siRNAs act as modulators of gene expression at the post-transcriptional level and are key players in stress response (Sunkar 2010). miRNAs regulate the expression of the target transcripts/genes by binding to reverse complementary sequences, causing cleavage of the target RNA, whereas siRNAs bind to the target sequence in a similar manner and direct DNA methylation (Khraiwesh et al. 2011). Several studies have been conducted to date that elucidate the involvement of miRNAs in various abiotic and biotic stresses including drought (Zhao et al. 2007; Liu et al. 2008; Zhou et al. 2010), cold (Zhou et al. 2008), salinity (Liu et al. 2008; Sunkar et al. 2008), UV-B radiation (Zhou et al. 2007), mechanical stress (Lu et al. 2005), and bacterial infection (Navarro et al. 2006) in various model plants. With the advancement in high-throughput sequencing and small RNA profiling techniques, along with the availability of advanced data analysis tools and software, the sequencing of a large number of small RNA libraries simultaneously and identification of stress-responsive miRNAs with precision have become faster and cheaper (Ding et al. 2013; Rajwanshi et al. 2014). However, only recently there have been a few reports on the identification of stress-responsive miRNAs from foxtail millet. A genome-wide transcriptome study in *S. italica* under drought stress was performed by Qi et al. (2013).

Two RNA and sRNA libraries were constructed from two treatments, namely drought stressed, and unstressed whole seedlings *S. italica* Yugu1 and sequenced. sRNAs of varied nucleotide sequence lengths were identified—24-nt (nucleotide) sRNAs were found to be predominant followed by 21-, 22-, and 23-nt sRNAs. The study also revealed that a number of 24-nt siRNAs were low across the genic regions, indicating their negative role in influencing gene expression in response to drought stress. Differential expression analysis led to the identification of 19 long maximally expressed noncoding RNAs during drought stress and, among these, two natural antisense transcripts (NATs of Si003758m and Si038715m) showed drought-regulated expression patterns (Qi et al. 2013). Similarly, Yi et al. (2013) constructed two small RNA libraries from shoot tissue of *S. italica* inbred line Yugu1 followed by sequencing using Illumina HighSeq 2000 platform leading to the identification of a total 43 known miRNAs, 172 novel miRNAs, and 2 miRNA precursor candidates. The targets of the selected miRNAs were identified, annotated, and functionally validated by stem-loop RT-PCR in four tissues (Yi et al. 2013). The raw reads generated from the studies of Yi et al. (2013) and Qi et al. (2013) are available in the NCBI SRA database under accession numbers SRA062640 and SRA062827, respectively.

Han et al. (2014) analyzed expression patterns of 43 miRNAs in different tissues of *S. italica*, including leaves, roots, stems, and spikes, and also experimentally validated five predicted targets of four miRNAs using 5′-RLM-RACE. Furthermore, a total of 355 mature miRNAs (Sit-miR) have been identified from the genomic and CDS sequences of *S. italica* and classified into 53 families (Khan et al. 2014). Expression profiling of candidate Sit-miRs was analyzed using Northern blot analysis and stem-loop RT-qPCR under different abiotic stresses in two *S. italica* cultivars (IC-403579, stress tolerant; IC-480117, stress susceptible). Three Sit-miRs, namely Sit-miR162a, Sit-miR167b, and Sit-miR171b, were found to be up-regulated in tolerant cultivar compared to sensitive and Sit-miR156c, Sit-miR397a, Sit-miR393, Sit-miR160d, and Sit-miR6248a were found to be down-regulated in the tolerant cultivar under drought stress (Khan et al. 2014). The complete information about chromosomal location, length, sequences of pre-miRNA and mature miRNA, secondary structure, and target gene information of identified sit-miRs has been made available to the global research community through an open-access web resource, Foxtail millet miRNA Database (http://59.163.192.91/FmMiRNADb/ index.html; Khan et al. 2014).

In another study, four small RNA libraries from control and drought stressed seedlings of IC-403579 (tolerant) and IC-480117 (susceptible) were constructed and sequenced using Illumina HiSeq 2000 platform (Yadav et al. 2016). Transcriptome results of this study led to the identification of a total of 55 known miRNAs, which were classified into 23 miRNA families and 136 novel miRNAs which could be classified into 47 miRNA families, respectively. Stem-loop quantitative real-time PCR further validated some of these candidate novel dehydration-responsive Sit-miRs. In a similar kind of study, drought-responsive miRNAs were identified from *S. italica* inbred An04-4783 seedlings using Illumina sequencing (Wang et al. 2016). A total of 81 known and 76 novel miRNAs were identified. Furthermore, degradome sequencing was carried out to confirm the target genes of the identified drought-responsive miRNAs. The above studies give an insight into the role of miRNAome of *S. italica* in response to abiotic stress. There is also a dire need for identification and characterization of target genes of sRNAs to delineate their roles in abiotic stress regulatory pathways in foxtail millet.

8.6 DNA Methylation

Alteration of gene expression during plant development and various environmental stresses may also occur because of epigenetic modifications such as DNA methylation, chromatin remodeling, and histone modification (Boyko and Kovalchuk 2008). Many cytosines in the genome, particularly those located at CG dinucleotide, may potentially act as transcriptional switches (Shibuya et al. 2009). Hypomethylation or hypermethylation of DNA may function as a major switch to control the expression of effector genes during stress response in plants (Chinuusamy and Zhu 2009). Very little is known about the global patterns of epigenetic modification such as DNA methylation in the case of abiotic stress, and whether DNA methylation plays a role in various abiotic stress responses in plants is still unclear. Considering this, a genome-wide salinity-induced differential methylation analysis was carried out on 3-weeks-old contrasting foxtail millet cultivars, namely salinity tolerant cv. IC403579 (IC04) and sensitive cv. IC480117 (IC41) using a methylation-sensitive amplified polymorphism (MSAP) technique. The study indicated a reduction in DNA methylation level in the tolerant cultivar compared to the sensitive. It also led to the identification of a total of 86 MSAP fragments, which were functionally annotated and classified into various categories, namely transporters, TFs, phosphatases, oxidoreductases, transposable elements (TEs), etc. A validation of methylome data was also carried out through expression analysis of four selected genes. Overall, the study inferred the role of salinity stress in inducing genome-wide DNA methylation as well as gene expression modulation in foxtail millet (Pandey et al. 2017).

8.7 Advances in Foxtail Millet Genomic Resources and Their Utilization

Although foxtail millet is an important grain crop and a potential model for several bioenergy crops, the availability of genomic tools is limited and needs to be developed. ESTs and molecular markers can be excellent genomic resources for foxtail millet. Zhang et al. (2007b) analyzed the dehydration-induced transcriptome of foxtail millet cv. Mar51 using subtracted cDNA library and microarray. A total of 1,947 uniESTs were obtained of which 95 and 57 ESTs were up-regulated in roots and shoots, respectively. Jayaraman et al. (2008) used cDNA-AFLP markers for comparing the gene expression profiles of a salt tolerant and a salt sensitive cultivar of foxtail millet (*S. italica*) in response to salt stress. A total of 90 differentially expressed transcript-derived fragments (TDFs) were identified, out of which 86 TDFs were classified on the basis of their either complete presence or absence (qualitative variants) and 4 on differential expression pattern levels (quantitative variants) in the two varieties. Other than these, the genes identified in transcript profiling studies on drought (Lata et al. 2010) and salinity (Puranik et al. 2011a) are important genomic resources, which could be exploited for enhancing drought tolerance traits and can be functionally validated to understand the molecular genetics of foxtail millet in stress response and adaptation. Recently, Qi et al. (2013) analyzed the whole transcriptome of foxtail millet using the Illumina platform and identified a total of 2,484 drought responsive genes, out of which $\sim$48% were up-regulated and $\sim$52% were down-regulated. Sequence-based molecular markers, genetic linkage maps, and trait-genetics are important molecular tools in marker-aided selection and breeding for desired agronomic traits such as yield, quality, and improved abiotic stress tolerance. The availability of the foxtail millet genome sequence in the public domain has been an immense help in the development of large-scale genomic resources, comparative mapping, markers/quantitative trait loci (QTLs) identification and development, and molecular breeding of foxtail millet and related cereals crops (Muthamilarasan and Prasad 2015). The first assembled reference genome of foxtail millet and green foxtail was released independently by the United States Department of Energy-Joint Genome Initiative (USDOE-JGI) and Beijing Genome Initiative (BGI), China in 2012 (Bennetzen et al. 2012; Zhang et al. 2012; reviewed in Lata and Prasad 2013). Both studies predicted the genome size of foxtail millet to be $\sim$490 Mb coding for $\sim$30,000 genes. The studies also led to the identification of large number of stress-responsive genes and miRNA families in foxtail millet. In addition, thousands of SNPs, insertion-deletion polymorphisms, and structural variations were also identified in the sequenced genomes when compared with other lines.

In light of the availability of the genome sequence, comparative genomics studies have been geared up for the identification of genes underlying QTL(s) for determining key traits for genetic improvement of foxtail millet (Paterson et al. 1995; Doust and Kellogg 2006; Doust et al. 2009, 2010; Lata et al. 2013; Muthamilarasan and Prasad 2015). Foxtail millet genome sequence was also helpful in aligning the switchgrass genome (http://kdbioinfo.cropsoil.uga.edu/devoslab/prjctfxtlmllt.html), leading to translating the sequence and QTL information from this diploid crop to other candidate biofuel grasses, which are otherwise polyploid. A direct practical proposition of using foxtail millet genome sequence would therefore be an easy access for crop breeders searching for exploitable variations for crop improvement strategies. The genome sequence information has also served as a reference in whole-genome resequencing (WGR) of 916 *S. italica* accessions collected from different eco-geographical zones of the world and the construction of a high-density haplotype map using 85 million single nucleotide polymorphisms, which revealed genomic variations among these accessions (Jia et al. 2013). Similarly, the *S. italica* genome sequence data facilitate WGR of cultivated and wild varieties of *Setaria* with contrasting phenotypes to identify novel genes/alleles/QTLs underlying drought response and to execute NGS-based genomics-assisted breeding for drought tolerance. Although the *S. viridis* genome has also been sequenced (Bennetzen et al. 2012), the lack of publicly available sequence information has for a long time significantly impeded the development of genetic and genomic resources in this important model species. However, recent developments have made this data available in the web portal "*Setariabase*" (http://www.sviridis. org; Brutnell et al. 2015).

Sequence-based molecular markers, QTLs, and genetic linkage maps can also be effectively utilized for marker-aided breeding of economically important traits. Lata et al. (2011a) have identified an SNP in the *SiDREB2* gene of foxtail millet associated with dehydration tolerance. An ASM has also been developed for the same, which can amplify the specific products from the dehydration-tolerant and sensitive accessions (Lata et al. 2011a). It has also been shown that this ASM contributes to approximately one-fourth of the total variation in lipid peroxidation and relative water content at the seedling stage (Lata and Prasad 2013b, 2014). Furthermore, considering the importance of intron length polymorphic (ILP) markers in germplasm characterization, diversity studies, and molecular breeding, a set of 98 ILP markers was developed from foxtail millet (Gupta et al. 2011). In continuation of this, 5,123 ILP markers were also recently developed, and their utility in cross-species transferability, germplasm characterization, and comparative mapping among millets and non-millets was also demonstrated (Muthamilarasan et al. 2013). Foxtail millet has very limited resources of co-dominant microsatellite markers and saturated genetic linkage maps. In light of this, Pandey et al. (2013) performed a genome-wide analysis and identified 28,342 microsatellite repeat motifs spanning 405.3 Mb of foxtail millet genome. Among these microsatellites, trinucleotide repeats ($\sim$48%) were predominant followed by dinucleotide repeats ($\sim$75%). About 159 markers were validated successfully in 8 accessions of *Setaria sp.* with $\sim$67% polymorphic potential for different traits. Similarly, Kumari et al. (2013) identified 495 EST-SSRs from 66,027 ESTs available at the NCBI database and developed 447 SSR markers successfully validated in green foxtail and other millets. In another study, Zhang et al. (2014) isolated 5,020 highly repetitive microsatellite motifs from Yugu1 genome and designed 788 SSR primer pairs based on the sequence comparison between *S. italica* and *S. viridis*. Out of these, 733 could produce reproducible amplicons and were found to be polymorphic for 28 *Setaria* genotypes. Furthermore, Yadav et al. (2014) developed 176 miRNA-based molecular markers. In a similar line, the highly polymorphic nature of 20,278 TE-based markers were developed which could be categorized into five different polymorphism types, namely retrotransposon-based insertion polymorphisms, inter-retrotransposon amplified polymorphisms, repeat junction markers, repeat junction–junction markers, insertion-site-based polymorphisms and retrotransposon-microsatellite amplified polymorphisms (Yadav et al. 2015). Open access databases (Foxtail millet Transposable Elements-based Marker Database; FmTEMDb; http://59.163.192.83/ltrdb/index.html; and Foxtail Millet microRNA Database; FmMiRNADb; http://59.163.192.91/FmMiRNADb/) for these resources were also constructed for open use by the researchers. The details of molecular markers and genetic linkage maps reported to date in foxtail millet are listed in Table 8.3. Furthermore, the ASM, microsatellites, and other molecular markers would assist and play a significant role in allele mining of germplasm resources, diversity and transferability studies, phylogenetics, and comparative mapping, thus paving the way for discovery and exploitation of novel alleles in crop improvement, and accelerating the foxtail millet breeding process for stress tolerance.

8.8 Conclusion and Future Perspectives

Considering the potential of both foxtail millet and green foxtail as an excellent model system for evolutionary, architectural and physiological studies of related bioenergy grasses and C_4 crops, much attention has been given to these crops in the last decade by the scientific community. In fact, foxtail millet has gained considerable attention in terms of both structural and functional genomics. Furthermore, the release of the *Setaria* genome sequence has paved the way for large-scale development of genomic resources for this important model crop. Notably, foxtail millet research pertaining to various abiotic stresses has led to the identification of several potential candidate genes that could be effectively utilized in crop improvement programs including transgenics and molecular breeding. Furthermore, the development of various databases would act as a valuable resource for large-scale genotyping applications in foxtail millet. Despite genome-wide investigation of several gene families, there is still much scope for further studies that could help to provide a better understanding of the complex abiotic stress

Table 8.3 Summary of DNA-based markers and QTLs available in foxtail millet related to abiotic stress tolerance

DNA markers/QTLs	References
100 polymorphic SSRs developed from the two genomic DNA libraries	Jia et al. (2009)
~1000 SNPs by sequencing pools of RILs (*S. italica* acc. B100 x *S. virdis* acc. A10)	http://www.plantbio.uga.edu/media/2010_grad_symposium(1).pdf
98 ILP markers	Gupta et al. (2011)
One Allele-specific marker for *SiDREB2* gene	Lata et al. (2011a)
147 genomic microsatellite markers	Gupta et al. (2011)
28 342 microsatellites	Pandey et al. (2013)
447 EST-SSRs	Kumari et al. (2013)
788 SSRs	Zhang et al. (2014)
5,123 ILP markers	Muthamilarasan et al. (2014a, b)
176 miRNA-based markers	Yadav et al. (2014)
20,278 TE-based markers	Yadav et al. (2015)
8 QTLs and 128 SSR markers spans 1293.9 cm with an average of 14 markers per linkage group of the 9 linkage groups	Qie et al. (2014)

regulatory mechanisms operating in foxtail millet. The role of small RNAs and epigenetic modifications in abiotic stress response remains elusive, and therefore concerted efforts should be directed toward understanding their roles in abiotic stress response and adaptation. Further whole genome resequencing of diverse foxtail millet accessions would help not only in discovering novel variants for various abiotic stresses but also in exploiting them for the development of molecular markers and identification of QTLs to be utilized in foxtail millet breeding programs. Taken together, the potential abiotic stress tolerance features and the development of large-scale genetic resources in foxtail millet would certainly help in accelerating the crop improvement programs in this climate resilient crop.

Acknowledgements Charu Lata acknowledges INSPIRE Faculty Award [IFA-11LSPA-01] from Department of Science & Technology (DST), GoI, New Delhi. She is also thankful to the Director, CSIR-National Botanical Research Institute, Lucknow, India for providing facilities and support to conduct research in millet genomics.

References

Amadou I, Amza T, Shi Y-H, Le G-W (2011) Chemical analysis and antioxidant properties of foxtail millet bran extracts. Songklanakarin J Sci Technol 33:509–515

Ambawat S, Sharma P, Yadav NR, Yadav RC (2013) MYB transcription factor genes as regulators for plant responses: an overview. Physiol Mol Biol Plants 19:307–321

Baltensperger DD (2002) Progress with proso, pearl, and other millets. In: Janick J, Whipley A (eds) Trends in new crops and new uses. ASHS Press, Alexandria, pp 100–103

Barton L, Newsome SD, Chen FH, Wang H, Guilderson TP, Bettinger RL (2009) Agricultural origins and the isotopic identity of domestication in northern China. Proc Natl Acad Sci 106:5523–5528

Benabdelmouna A, Darmency H (2003) Copia -like retrotransposons in the genus *Setaria*: Sequence heterogeneity, species distribution and chromosomal organization. Plant Syst Evol 237:127–136

Benabdelmouna A, Abirached-Darmency M, Darmency H (2001) Phylogenetic and genomic relationships in *Setaria italica* and its close relatives based on the molecular diversity and chromosomal organization of 5S and 18S-5.8S-25S rDNA genes. Theor Appl Genet 103:668–677

Bennetzen JL, Schmutz J, Wang H, Percifield R, Hawkins J, Pontaroli AC, Estep M et al (2012) Reference genome sequence of the model plant *Setaria*. Nature Biotechnol 30:555–561

Bonthala VS, Muthamilarasan M, Roy R, Prasad M (2014) FmTFDb: a foxtail millet transcription factors database for expediting functional genomics in millets. Mol Biol Rep 41:6343–6348

Boyko A, Kovalchuk I (2008) Epigenetic control of plant stress response. Environ Mol Mutagen 49:61–72

Brutnell TP, Bennetzen JL, Vogel JP (2015) *Brachypodium distachyon* and *Setaria viridis*: model genetic systems for the grasses. Annu Rev Plant Biol 66:465–485

Chen Z, Chen M, Xu ZS, Li LC, Chen XP, Ma YZ (2014) Characteristics and expression patterns of the aldehyde dehydrogenase (ALDH) gene superfamily of foxtail millet (*Setaria italica L.*). PLoS One 9(7):e101136

Chen J, Qi Y (1993) Recent developments in foxtail millet cultivation and research in China. In: Riley KW, Gupta SC, Seetharam A, Mushonga JN (eds) Advances in small millets. Oxford and IBH Publishing Co, New Delhi, pp 101–107

Chinnusamy V, Zhu JK (2009) Epigenetic regulation of stress responses in plants. Curr Opin Plant Biol 12:133–139

Ding Y, Tao Y, Zhu C (2013) Emerging roles of microRNAs in the mediation of drought stress response in plants. J Exp Bot 64:3077–3086

Doust AN, Kellogg EA (2006) Effect of genotype and environment on branching in weedy green millet (*Setaria viridis*) and domesticated foxtail millet (*Setaria italica*) (Poaceae). Mol Ecol 15:1335–1349

Doust AN, Kellogg EA, Devos KM, Bennetzen JL (2009) Foxtail millet: a sequence-driven grass model system. Plant Physiol 149:137–141

Doust AN, Mauro-Herrera M, Malahy M, Stromski J, Estep M, Percifield R, Wang H, Wu L, Wu X, Zale J, Devos K, Bennetzen J (2010) Development of genomic and genetic tools for foxtail millet, and use of these tools in the improvement of biomass production for bioenergy crops. In: Plant and animal genomes XVIII conference. 9–13 January, San Diego, CA. P372 (abstract)

Dvořáková Z, Čepková PH, Janovská D, Viehmannová I, Svobodová E, Cusimamani EF, Milella L (2015) Comparative analysis of genetic diversity of 8 millet genera revealed by ISSR markers. Emirates J Food Agric 27:617–628

Dwivedi S, Upadhyaya H, Senthilvel S, Hash C, Fukunaga K, Diao X, Santra D, Baltensperger D, Prasad M (2012). In: Janick J (ed) Millets: genetic and genomic resources. Plant Breed Rev, vol 35. Wiley, USA, pp 247–375

Feng ZJ, He GH, Zheng WJ, Lu PP, Chen M, Gong YM, Ma YZ, Xu ZS (2015) Foxtail millet NF-Y families: genome-wide survey and evolution analyses identified two functional genes important in abiotic stresses. Front Plant Sci 6:1142

Feng ZJ, Xu ZS, Sun J, Li LC, Chen M, Yang GX, He GY, Ma YZ (2016) Investigation of the ASR family in foxtail millet and the role of ASR1 in drought/oxidative stress tolerance. Plant Cell Rep 35:115–128

Goron TL, Raizada MN (2015) Genetic diversity and genomic resources available for the small millet crops to accelerate a new green revolution. Front Plant Sci 6:157. doi:10.3389/fpls.2015.00157

Gupta S, Kumari K, Das J, Lata C, Puranik S, Prasad M (2011) Development and utilization of novel intron length polymorphic markers in foxtail millet (*(L.) P. Beauv.*). Genome 54(7):586–602

Gupta S, Kumari K, Muthamilarasan M, Parida SK, Prasad M (2014) Population structure and association mapping of yield contributing agronomic traits in foxtail millet. Plant Cell Rep 33:881–893

Han J, Xie H, Sun Q, Wang J, Lu M, Wang W, Guo E, Pan J (2014) Bioinformatic identification and experimental validation of miRNAs from foxtail millet (*Setaria italica*). Gene 546(2):367–377

He Z, Bonjean APA (2010) Cereals in China. CIMMYT, Mexico

Huang Z, Zhao L, Chen D, Liang M, Liu Z, Shao H et al (2013) Salt stress encourages proline accumulation by regulating proline biosynthesis and degradation in Jerusalem Artichoke Plantlets. PLoS ONE 8(4): e62085

Huang P, Feldman M, Schroder S, Bahri BA, Diao X, Zhi H, Estep M, Baxter I, Devos KM, Kellogg EA (2014) Population genetics of Setaria viridis, a new model system. Mol Ecol 23:4912–4925

Jacinto T, Farmer EE, Ryan CA (1993) Purification of potato leaf plasma membrane protein pp34, a protein phosphorylated in response to oligogalacturonide signals for defense and development. Plant Physiol 103:1393–1397

Jayaraman A, Puranik S, Rai NK, Vidapu S, Sahu PP, Lata C et al (2008) cDNA-AFLP analysis reveals differential gene expression in response to salt stress in foxtail millet (*Setaria italica* L.). Mol Biotechnol 40:241–251

Jia G, Huang X, Zhi H, Zhao Y, Zhao Q, Li W, Chai Y, Yang L, Liu K et al (2013) A haplotype map of genomic variations and genome-wide association studies of agronomic traits in foxtail millet (*Setaria italica*). Nat Genet 45(8):957–961

Khan Y, Yadav A, Bonthala VS, Muthamilarasan M, Yadav CB, Prasad M (2014) Comprehensive genome-wide identification and expression profiling of foxtail millet [*Setaria italica (L.)*] miRNAs in response to abiotic stress and development of miRNA database. Plant Cell Tiss Org Cult (PCTOC) 118 (2):279–292

Khraiwesh B, Zhu J-K, Zhu J (2011) Role of miRNAs and siRNAs in biotic and abiotic stress responses of plants. Biochem Biophys Acta. doi:10.1016/j.bbagrm. 2011.05.001

Krishnamurthy L, Serraj R, Rai KN, Hash CT, Dakheel AJ (2007) Identification of pearl millet [*Pennisetum glaucum (L.) R.Br.*] lines tolerant to soil salinity. Euphytica 158(1–2):179–188

Kumari K, Muthamilarasan M, Misra G, Gupta S, Subramanian A, Parida SK, Chattopadhyay D, Prasad M, Jordan IK (2013) Development of eSSR-Markers in *Setaria italica* and their applicability in studying genetic diversity, cross-transferability and comparative mapping in millet and non-millet Species. PLoS ONE 8(6):e67742

Lata C, Prasad M (2011) Role of DREBs in regulation of abiotic stress responses in plants. J Exp Bot 14:4731–4748

Lata C, Prasad M (2013a) *Setaria* genome sequencing: an overview. J Plant Biochem Biotechnol 22:257–260

Lata C, Prasad M (2013b) Validation of an allele-specific marker associated with dehydration stress tolerance in a core set of foxtail millet accessions. Plant Breed 132:496–499

Lata C, Prasad M (2014) Association of an allele-specifi c marker with dehydration stress tolerance in foxtail millet suggests *SiDREB2* to be an important QTL. J Plant Biochem Biotechnol 23:119–122

Lata C, Sahu PP, Prasad M (2010) Comparative transcriptome analysis of differentially expressed genes in foxtail millet (*Setaria italica* L.) during dehydration stress. Biochem Biophys Res Commun 393:720–727

Lata C, Bhutty S, Bahadur RP, Majee M, Prasad M (2011a) Association of an SNP in a novel DREB2-like gene SiDREB2 with stress tolerance in foxtail millet [*Setaria italica* (L.)]. J Exp Bot 62:3387–3401

Lata C, Jha S, Dixit V, Sreenivasulu N, Prasad M (2011b) Differential antioxidative responses to dehydration-induced oxidative stress in core set of foxtail millet cultivars [*Setaria italica* (L.)]. Protoplasma 248:817–828

Lata C, Gupta S, Prasad M (2013) Foxtail millet: a model crop for genetic and genomic studies in bioenergy grasses. Crit Rev Biotechnol 33:328–343

Lata C, Mishra AK, Muthamilarasan M, Bonthala VS, Khan Y, Prasad M (2014) Genome-wide investigation and expression profiling of AP2/ERF transcription factor superfamily in foxtail millet (*Setaria italica* L.). PLoS ONE 9:e113092

Layton DJ, Kellogg EA (2014) Morphological, phylogenetic, and ecological diversity of the new model species *Setaria viridis* (Poaceae: Paniaceae) and its close relatives. Am J Bot 101:539–557

Li Y-M (1991) A study on the identification of drought-resistance on millet germplasm (in Chinese. English abstract). Acta Agril Boreali-Sinica 6:20–25

Li Y-M (1997) Breeding for foxtail millet drought tolerant cultivars (in Chinese). In: Li Y (ed) Foxtail millet breeding. Chinese Agr Press, Beijing, pp 421–446

Li P, Brutnell TP (2011) *Setaria viridis* and *Setaria italica*, model genetic systems for the Panicoid grasses. J Exp Bot 62:3031–3037

Li Y, Wu SZ (1996) Traditional maintenance and multiplication of foxtail millet (*Setaria italica*

(L) P. Beauv.) landraces in China. Euphytica 87:33–38

Li C, Yue J, Wu X, Xu C, Yu J (2014) An ABA-responsive DRE-binding protein gene from *Setaria italica*, *SiARDP*, the target gene of *SiAREB*, plays a critical role under drought stress. J Exp Bot 65:5415–5427

Li J, Dong Y, Li C, Pan Y, Yu J (2017) *SiASR4*, the target gene of *SiARDP* from *Setaria italica*, improves abiotic stress adaption in plants. Front Plant Sci 7:2053

Liang S, Yang G, Ma Y (2010) Chemical characteristics and fatty acid profile of foxtail millet bran oil. J Am Oil Chem Soc 87:63–67

Liu HH, Tian X, Li YJ, Wu CA, Zheng CC (2008) Microarray-based analysis of stress-regulated microRNAs in *Arabidopsis thaliana*. RNA 14:836–843

Lu S, Sun YH, Shi R, Clark C, Li L, Chiang VL (2005) Novel and mechanical stress-responsive microRNAs in *Populus trichocarpa* that are absent from Arabidopsis. Plant Cell 17:2186–2203

Mirzaee MR, Zare R, Nasrabad AA (2010) A new leaf and sheath brown spot of foxtail millet caused by *Bipolaris australiensis*. Australasian Plant Disease Notes 5:19–20

Mishra AK, Puranik S, Bahadur RP, Prasad M (2012) The DNA binding activity of an AP2 protein is involved in transcriptional regulation of a stress-responsive gene, SiWD40, in foxtail millet. Genomics. doi:10.1016/j.ygeno.2012.06.012

Munirathnam P, Venkatramanamma K, Anusha A (2015) valuation of foxtail millet genotype s for blast and rust diseases under field conditions. Current Biotica 9:263–268

Muthamilarasan M, Prasad M (2015) Advances in Setaria genomics for genetic improvement of cereals and bioenergy grasses. Theor Appl Genet 128:1–14

Muthamilarasan M, Venkata Suresh B, Pandey G, Kumari K, Parida SK, Prasad M (2013) Development of 5123 intron-length polymorphic markers for large-scale genotyping applications in foxtail millet. DNA Res 21:41–52

Muthamilarasan M, Bonthala VS, Mishra AK, Khandelwal R, Khan Y, Roy R, Prasad M (2014a) C2H2-type of zinc finger transcription factors in foxtail millet define response to abiotic stresses. Funct Integr Genomics 14:531–543

Muthamilarasan M, Khandelwal R, Yadav CB, Bonthala VS, Khan Y, Prasad M (2014b) Identification and molecular characterization of MYB transcription factor superfamily in C4 model plant foxtail millet (*Setaria italica* L.). PLoS ONE 9:e109920

Muthamilarasan M, Bonthala VS, Khandelwal R, Jaishankar J, Shweta S, Nawaz K, Prasad M (2015) Global analysis of WRKY transcription factor superfamily in *Setaria* identifies potential candidates involved in abiotic stress signaling. Front Plant Sci 6:910

Nakayama H, Nagamine T, Hayashi N (2005) Genetic variation of blast resistance in foxtail millet (*Setaria italica* (L.) P. Beauv.) and its geographic distribution. Genet Resour Crop Evol 52:863–868

Navarro L, Dunoyer P, Jay F, Arnold B, Dharmasiri N, Estelle M, Voinnet O, Jones JD (2006) A plant miRNA contributes to antibacterial resistance by repressing auxin signaling. Science 312:436–439

Pan Y, Li J, Jiao L, Li C, Zhu D, Yu J (2016) A Non-specific *Setaria italica* lipid transfer protein gene plays a critical role under abiotic stress. Front Plant Sci 7

Pandey G, Misra G, Kumari K, Gupta S, Parida SK, Chattopadhyay D, Prasad M (2013) Genome-wide development and use of microsatellite markers for large-scale genotyping applications in foxtail millet [*Setaria italica* (L.)]. DNA Res 20:197–207

Pandey G, Yadav CB, Sahu PP, Muthamilarasan M, Prasad M (2017) Salinity induced differential methylation patterns in contrasting cultivars of foxtail millet (*Setaria italica* L.). Plant Cell Rep 36:759–772

Paterson AH, Lin YR, Li Z, Schertz KF, Doebley JF, Pinson SRM, Liu SC, Stansel JW, Irvine JE (1995) Convergent domesticationof cereal crops by independent mutations at corresponding Genetic Loci. Science 269(5231):1714–1718

Peng Y, Zhang J, Cao G, Xie Y, Liu X, Lu M et al (2010) Overexpression of a PLDalpha1 gene from *Setaria italica* enhances the sensitivity of Arabidopsis to abscisic acid and improves its drought tolerance. Plant Cell Rep 29:793–802

Puranik S, Jha S, Srivastava PS, Sreenivasulu N, Prasad M (2011a) Comparative transcriptome analysis of contrasting foxtail millet cultivars in response to short-term salinity stress. J Plant Physiol 168:280–287

Puranik S, Bahadur RP, Srivastava PS, Prasad M (2011b) Molecular cloning and characterization of a membrane associated NAC family gene, SiNAC from foxtail millet [*Setaria italica* (L.) P. Beauv.]. Mol Biotechnol doi:10.1007/s12033-011-9385-7

Puranik S, Sahu PP, Mandal SN, Venkata Suresh B, Parida SK, Prasad M (2013) Comprehensive genome-wide survey, genomic constitution and expression profiling of the NAC transcription factor family in foxtail millet (*Setaria italica* L.). PLoS ONE 8:e64594

Qi X, Xie S, Liu Y, Yi F, Yu J (2013) Genome-wide annotation of genes and noncoding RNAs of foxtail millet in response to simulated drought stress by deep sequencing. Plant Molec Biol 83(4-5):459–473

Qie L, Jia G, Zhang W, Schnable J, Shang Z, Li W, Liu B, Li M, Chai Y, Zhi H, Diao X (2014) Mapping of quantitative trait locus (QTLs) that contribute to germination and early seedling drought tolerance in the interspecific cross *Setaria italica*×*Setaria viridis*. PLoS One 9(7):e101868. doi: 10.1371/journal.pone.0101868

Rajwanshi R, Chakraborty S, Jayanandi K, Deb B, Lightfoot DA (2014) Orthologous plant microRNAs: microregulators with great potential for improving stress tolerance in plants. Theor Appl Genet 127:2525–2543

Shantz HL, Piemeisel LN (1927) The water requirement of plants at Akron Colorado. J Agri Res. 34:1093–1189

Sharma R, Girish AG, Upadhyaya HD, Humayun P, Babu TK, Rao VP, Thakur RP (2014) Identification of blast resistance in a core collection of foxtail millet germplasm. Plant Dis 98:519–524

Shibuya K, Fukushima S, Takatsuji H (2009) RNA-directed DNA methylation induces transcriptional activation in plants. Proc Natl Acad Sci USA 106:1660–1665

Siles M, Baltensperger DD, Nelson LA, Marcon A, Frickel GE (2001) Registration of five genetic marker stocks for foxtail millet. Crop Sci 41:2011–2012

Singh RK, Jaishankar J, Muthamilarasan M, Shweta S, Dangi A, Prasad M (2016) Genome-wide analysis of heat shock proteins in C4 model, foxtail millet identifies potential candidates for crop improvement under abiotic stress. Sci Rep 6:32641

Sreenivasulu N, Ramanjulu S, Ramachandra-Kini K, Prakash HS, Shekar-Shetty H, Savithri HS, Sudhakar C (1999) Total peroxidase activity and peroxidase isoforms as modified by salt stress in two cultivars of fox-tail millet with differential salt tolerance. Plant Sci 141:1–9

Sreenivasulu N, Miranda M, Prakash HS, Wobus U, Weschke W (2004) Transcriptome changes in foxtail millet genotypes at high salinity: identification and characterization of a PHGPX gene specifically upregulated by NaCl in a salt-tolerant line. J Plant Physiol 161:467–477

Sunkar R (2010) MicroRNAs with macro-effects on plant stress responses. Semin Cell Dev Biol 21:805–811

Sunkar R, Zhu JK (2004) Novel and stress-regulated microRNAs and other small RNAs from Arabidopsis. Plant Cell 16:2001–2019

Sunkar R, Zhou X, Zheng Y, Zhang W, Zhu JK (2008) Identification of novel and candidate miRNAs in rice by high throughput sequencing. BMC Plant Biol 2008 (8):25

Tadele Z (2016) Drought adaptation in millets. In: Shanker AK, Shanker C (eds) Abiotic and biotic stress in plants—recent advances and future perspectives. ISBN 978-953-51-2250-0. http://dx.doi.org/10.5772/61929

Tian BH, Wang J, Zhang L, Li Y, Wang S, Li H (2010) Assessment of resistance to lodging of landrace and improved cultivars in foxtail millet. Euphytica 172:295–302

Tian BH, Liu Y, Zhang LX, Song SX (2015) Characterization of culm morphology, anatomy and chemical composition of foxtail millet cultivars differing in lodging resistance. J Agril Sci 153:1437–1448. doi:10.1017/S0021859614001105

Veeranagamallaiah G, Chandraobulreddy P, Jyothsnakumari G, Sudhakar C (2007) Glutamine synthetase expression and pyrroline-5-carboxylate reductase activity influence proline accumulation in two cultivars of foxtail millet (*Setaria italica* L.) with differential salt sensitivity. Environ Expt Bot 60:239–244

Veeranagamallaiah G, Jyothsnakumari G, Thippeswamy M, Reddy PCO, Surabhi G-K, Sriranganayakulu G, Mahesh Y, Rajasekhar B, Madhurarekha C, Sudhakar C (2008) Proteomic analyses of salt responses in foxtail millet (*Setaria italica* L. cv. Prasad) seedlings. Plant Sci 175:631–641

Veeranagamallaiah G, Ranganayakulu GS, Thippeswamy M, Sivakumar M, Reddy EK, Pandurangaiah M, Sridevi V, Sudhakar C (2009) Aldose reductase expression contributes in sorbitol accumulation and 4-hydroxynon-2-enal detoxification in two foxtail millet (*Setaria italica* L.) cultivars with different salt stress tolerance. Plant Growth Regul 59:137–143

Wang RL, Wendel JF, Dekker JH (1995) Weedy adaptation in *Setaria* spp. I. Isozyme analysis of genetic diversity and population genetic structure in *Setaria viridis*. Am J Bot 82:308–317

Wang Y, Zhang J, Cui R, Li W, Zhi H, Li H, Diao X (2009) Transformation of wheat with DNAj gene from foxtail millet via pollen-tube pathway (in Chinese with English abstract) Acta Agril Boreali-Sinica 2009-02. (http://en.cnki.com.cn/Article_en/CJFDTOTALHBNB200902005.htm)

Wang C, Chen J, Zhi H, Yang L, Li W, Wang Y, Li H, Zhao B, Chen M, Diao X (2010) Population genetics of foxtail millet and its wild ancestor. BMC Genet 11:90

Wang M, Li P, Li C, Pan Y, Jiang X, Zhu D, Zhao Q, Yu J (2014a) SiLEA14, a novel atypical LEA protein, confers abiotic stress resistance in foxtail millet. BMC Plant Biol 14:290

Wang Y, Liu H, Xin Q (2014b) Genome-wide analysis and identification of cytokinin oxidase/dehydrogenase (CKX) gene family in foxtail millet (*Setaria italica*). Crop J 2:244–254

Wang Y, Li L, Tang S, Liu J, Zhang H, Zhi H, Jia G, Diao X (2016) Combined small RNA and degradome sequencing to identify miRNAs and their targets in response to drought in foxtail millet. BMC Genet 17:57

Wen Q-F, Wang L, Wang XY (2005) The foxtail millet germplasm resources and screening and utilization of drought resistance germplasm in Shanxi (In Chinese with English abstract). J Shanxi Agr Sci 33:32–33

Yadav CB, Muthamilarasan M, Pandey G, Khan Y, Prasad M (2014) Development of novel microRNA-based genetic markers in foxtail millet for genotyping applications in related grass species. Mol Breed. doi:10.1007/s11032-014-0137-9

Yadav CB, Bonthala VS, Muthamilarasan M, Pandey G, Khan Y, Prasad M (2015) Genome-wide development of transposable elements-based markers in foxtail millet and construction of an integrated database. DNA Res 22:79–90

Yadav A, Khan Y, Prasad M (2016) Dehydration-responsive miRNAs in foxtail millet: genome-wide identification, characterization and expression profiling. Planta 243:749–766

Yi F, Xie S, Liu Y, Qi X, Yu J (2013) Genome-wide characterization of microRNA in foxtail millet (*Setaria italica*). BMC Plant Biol 13(1):212

Yue J, Li C, Liu Y, Yu J (2014) A remorin gene SiREM6, the target gene of SiARDP, from foxtail millet (*Setaria italica*) promotes high salt tolerance in transgenic Arabidopsis. PLoS ONE 9(6):e100772

Zhang JP, Wang MY, Bai YF, Jia JP, Wang GY (2005) Rapid evaluation on drought tolerance of foxtail millet at seedling stage (in Chinese. English abstract). J Plant Genet Resour 6:59–62

Zhang C, Zhang H, Li JX (2007a) Advances of millet research on nutrition and application. J Chinese Cereals Oils Assoc 22:51–55

Zhang J, Liu T, Fu J, Zhu Y, Jia J, Zheng J, Zhao Y, Zhang Y, Wang G (2007b) Construction and application of EST library from *Setaria italica* in response to dehydration stress. Genomics 90:121–131

Zhang JP, Liu TS, Zheng J, Jin Z, Zhu Y, Guo JF, Wang GY (2007c) Cloning and characterization of a putative 12-oxophytodienoic acid reductase cDNA induced by osmotic stress in roots of foxtail millet. DNA Seq 18:138–144

Zhang G, Liu X, Quan Z, Cheng S, Xu X, Pan S, Xie M et al (2012) Genome sequence of foxtail millet (*Setaria italica*) provides insights into grass evolution and biofuel potential. Nature Biotechnol 30:549–554

Zhang S, Tang C, Zhao Q, Li J, Yang L, Qie L, Fan X et al (2014) Development of highly polymorphic simple sequence repeat markers using genome-wide microsatellite variant analysis in Foxtail millet [*Setaria italica* (L.) P. Beauv]. BMC Genom 15:78

Zhang L, Liu B, Zheng G, Zhang A, Li R (2017) Genome-wide characterization of the SiDof gene family in foxtail millet (*Setaria italica*). Biosystems 151:27–33

Zhao B, Liang R, Ge L, Li W, Xiao H, Lin H, Ruan K, Jin Y (2007) Identification of drought-induced microRNAs in rice. Biochem Biophys Res Commun 354:585–590

Zhao L, Zhao Q, Ao G, Yu J (2009) The foxtail millet Si69 gene is a Wali7 (wheat aluminium-induced protein 7) homologue and may function in aluminium tolerance. Chinese Sci Bullet 54:1697–1706

Zhi H, Diao X, Lu P, Li W, Akolova Z (2004) Methodology analysis on screening of salt tolerant genotypes from foxtail millet and other *Setaria* species. J Hebei Agr Res 8:15–18

Zhou X, Wang G, Zhang W (2007) UV-B responsive microRNA genes in *Arabidopsis thaliana*. Mol Syst Biol 3:103

Zhou X, Wang G, Sutoh K, Zhu JK, Zhang W (2008) Identification of cold-inducible microRNAs in plants by transcriptome analysis. Biochim Biophys Acta 1779:780–788

Zhou L, Liu Y, Liu Z, Kong D, Duan M, Luo L (2010) Genome-wide identification and analysis of drought-responsive microRNAs in *Oryza sativa*. J Exp Bot 61:4157–4168

Zhu X-H, Song Y-C, Zhao Z-H, Shi Y-S, Liu Y-H, Li Y, Wang T-Y (2008) Methods for identification of drought tolerance at germination period of foxtail millet by osmotic stress (in Chinese, English abstract). J Plant Genet Resour 9:62–67

Genetic Transformation of *Setaria*: A New Perspective

9

Priyanka Sood and Manoj Prasad

Abstract

Setaria italica and its wild progenitor *Setaria viridis*, collectively called *Setaria*, have recently emerged as promising translational research models for studying stress resistance, domestication, C_4 grass biology, and bioenergy traits. However, genetic engineering of *Setaria* remains challenging without the availability of robust transformation methods. Their recalcitrance to in vitro manipulation and transformation restricts the use of transgenesis and contemporary genome editing tools for crop functional genomics research. Not much work on transgenesis in *Setaria* has been reported. In the present chapter we aim at updating available information related to *Setaria* tissue culture and genetic transformation. In addition, we discuss different factors and methods to ease the transformation studies in *Setaria*. The advanced or alternative gene transformation methods with respect to *Setaria* are also discussed.

9.1 Introduction

Plant transformation technology has turned into a versatile platform that facilitates fundamental insights into plant biology. Every step of biotechnological application to crop improvement, starting from transgene regulation, studying gene function, dissecting biological processes, to developing improved varieties, relies on transformation technology. Despite technological advances, many crops still remain strenuous for regeneration and transformation. The success stories represent the conclusion of many years of venture in tissue culture improvement followed by transformation techniques and genetic engineering. Moreover, advancement of the genomics programme has led to foundational discoveries of crop genes and their functions that need to be translated to facilitate an increase in agricultural production. The major challenge of rapidly manipulating plant genomes through genetic transformation limits plant genetic and genome editing, thereby

P. Sood · M. Prasad (✉)
National Institute of Plant Genome Research (NIPGR), Aruna Asaf Ali Marg, New Delhi 110067, India
e-mail: manoj_prasad@nipgr.ac.in

M. Prasad (ed.), *The Foxtail Millet Genome*, Compendium of Plant Genomes,
DOI 10.1007/978-3-319-65617-5_9

hampering future plant breeding. This initiated concern about developing proficient plant transformation technologies and improving existing methodologies. The use of high-throughput and novel transformation systems could significantly enhance plant genomics research, thereby revolutionizing commercial agriculture. Similarly, genetic improvement of millets has been initiated in recent years despite their great importance.

Millets are annual C_4 grasses that belong to the family Poaceae (Gramineae) of the monocotyledon group. They rank sixth among the world's most important cereal grains. Although less prominent than major cereals, these small-seeded millets are grown worldwide for food, feed, forage, and fuel (O'Kennedy et al. 2006), sustaining more than one-third of the world's population (Changmei and Dorothy 2014). Millets are a major source of energy and protein for millions of people living in hot and dry areas of the world (Rachie 1975; Amadou et al. 2013; Muthamilarasan et al. 2016a). They include several annual food and fodder grasses such as pearl millet (*Pennisetum glaucum* (L.) R. Br), finger millet (*Eleusine coracana* (L.) Gaertn), foxtail millet (*Setaria italica* (L.) P. Beauvois), proso millet (*Panicum miliaceum* L.), barnyard millet (*Echinochloa crus-galli* L. Beauv.), kodo millet (*Paspalum scrobiculatum* L.), little millet (*Panicum sumatrense* Roth ex Roem. & Schult.), tef (*Eragrotis tef*), fonio (*Digitaria* sp.), guinea grass (*Panicum maximum* Jacq), elephant grass (*Pennisetum purpurium* Schumach.), and bahiagrass (*Paspalum notatum* Flugge) (Dwivedi et al. 2012).

Drought and lack of irrigation are enduring constraints to agricultural production (Ceccarelli and Grando 1996). In addition to drought, high temperature and soil salinity are other major factors that strongly affect growth and productivity of crops. Hence, securing enhanced yield stability totally relies on improving tolerance in crop plants against abiotic stresses. However, yield losses caused by abiotic constraints are high and unpredictable in nature with regard to duration, timing, and intensity of stresses. Because millets are quite resilient to drought, heat, and low nutrient input conditions (Saha et al. 2016), they are suitable for sustaining agriculture and food security on low fertility marginal lands of hilly and semi-arid regions of India that are not suitable for major cereal crops (Amadou et al. 2013). Despite their importance, millets are considered to be minor cereal crops of only regional significance and have remained an under-researched crop commodity. Negligible consideration is, thus, given to developing genomic and genetic resources for these crops. As global climate changes adversely influence the production of other major cereals, millets are substantially becoming attractive alternatives for small scale grain production (Tadele 2016). They also have the potential to meet increasingly rigorous weather patterns (Dai 2011) and other challenges. Although neglected previously, the value of millets toward agricultural stability in the new millennium has now begun to be recognized (Goron and Raizada 2015). Millets are thus, requisite plant genetic resources for the agriculture and food security of poor farmers residing in arid, uncultivable, and marginal lands. Fortunately, millet biodiversity provides the scientific community with prospective access to potential genes for crop improvement. Millets are, thus, valuable resources with a remarkable potential for the New Green Revolution in the era of a world with limiting natural resources and climate change.

9.2 Move Toward Foxtail Millet–Green Foxtail Pair as Model Grasses

Recently, millets such as *S. italica* L. (foxtail millet) and *Setaria viridis* (green foxtail), collectively called *Setaria*, gained promising scientific consideration as an attractive substitute for more established grass model systems such as rice, brachypodium, and maize (Muthamilarasan and Prasad 2015). *Setaria* can be genetically aligned to some of the world's most important food, feed, and energy crops (Defelice 2002; Doust et al. 2009; Li and Brutnell 2011; Brutnell 2015). Grasses of the genus *Setaria* belongs to the tribe Paniceae of the subfamily Pancoideae and are, as are all millet species, a member of the PACMAD clade. The Panicoideae subfamily is one of the biggest in the grasses, comprising many important crop species, namely maize (*Zea mays*), sugarcane (*Saccharum officinarum*), sorghum (*Sorghum bicolor*), foxtail millet (*S. italica*), common millet (*Panicum miliaceum*), pearl millet (*Pennisetum glaucum*), switchgrass (*Panicum virgatum*), napiergrass (*Pennisetum purpureum*), etc. Being part of this subfamily, with close evolutionary relationships, the *Setaria* model facilitates functional genomics for Panicoideae crops.

Genus *Setaria* is comprised of approximately 150 species distributed all over the world. Of these, foxtail millet was domesticated around 8,700 years ago in Northern China from *S. viridis* that is considered to be its wild progenitor. Previous studies illustrate *S. viridis* to be most closely related and inter-fertile with foxtail millet (Benabdelmouna et al. 2001; Huang et al. 2014). Close phylogenetic relatedness in general means sharing common genetic mechanisms driving the complex traits, thereby making it easier to translate genetic discoveries between closely related species. Foxtail millet is considered one of the world's oldest crops and ranks second in world total millet production (Yang et al. 2012). Although, maize and sorghum were initially considered as a model for C_4 functional genomics study, their large genome size, paleopolyploid evolution history, adult plant size, and longer life cycle present challenges for functional genomics research. Similar to maize, green foxtail and foxtail millet are C_4 plants, but they have smaller genomes and are true diploids. Likewise, polyploid backgrounds of putative biofuel crops such as switchgrass and napiergrass make them unsuitable for functional genomics study, similar to the situation for common millet, an ancient cereal. The establishment of *Setaria* as a model can assist in the genetic study of all these species, and many more Panicoid grasses. Their phylogenetic relatedness to several other millets and cereals, namely maize, sugarcane, sorghum, pearl millet, common millet, etc., gives them an additional advantage. The morphological and developmental similarity of *Setaria* to several important outbreeding, polyploids with larger genomes allows it to serve as a tractable diploid model for these valuable dry-land crops.

Together, they have specific characteristic advantages including small genomes ($\sim$500 Mb), short life cycle, simple diploid genetics, small stature, inbreeding nature, prolific seed production, and simple growth requirements that are not inherent to the majority of the millet species (Doust et al. 2009). These characteristics, along with the C_4 photosynthetic traits, genetic close-relatedness to major cereals and bioenergy grasses, and potential abiotic stress tolerance, have accentuated these crops as a tractable model for studying C_4 photosynthesis, abiotic stress tolerance, and biofuel traits (Diao et al. 2014). Considering the importance of *Setaria*, their genome has been sequenced and two genome sequences of foxtail millet were released in 2012 (Bennetzen et al. 2012; Zhang et al. 2012), followed by a third genome sequence of the landrace Daqingjie of foxtail millet that came a year later (Jia et al. 2013). The genome sequence of green foxtail Accession N10 was also released by the same group (Jia et al. 2013). Bai et al. (2013) re-sequenced the landrace Shilixiang of foxtail millet to facilitate identification of useful markers of agronomic importance. Therefore, with the availability of high-quality reference genome sequences, and a high-density haplotype map of genome variation, information in public databases has accelerated the development of genetic and genomic resources in these model crops. Mutant populations have also been characterized in both green foxtail and foxtail millet (Liu et al. 2016; Xue et al. 2016).

Researchers can now compare these C_4 temperate, cold- and drought-tolerant model grasses to other plants that may or may not have evolved these adaptations, particularly significant for biotechnological improvement. *Setaria* being a model is not only relevant to translating QTLs, mapped genes, and high-throughput phenotyping strategies to other millets, or for validating candidate genes. Rather, specific characteristic attributes of *Setaria* enhance its progress as a study system for biological processes, hypothesis testing, and genome engineering, with congruity to the entire plant kingdom and beyond. Together, studies have now successfully demonstrated utilization of the *S. italica* and *S. viridis* pair for (1) characterization of agronomic and domestication-related traits (Fang et al. 2016; Hodge and Kellogg 2016; Liu et al. 2016; Mauro-Herrera and Doust 2016), (2) fundamental research pertaining to biological processes and plant–microbe interactions (Li et al. 2015), and (3) translational research related to bioenergy traits, abiotic and biotic stress resistance pathways, nutritional pathways, and evolution of C_4 versus C_3 traits (Mandadi et al. 2014; Brutnell et al. 2015; Martin et al. 2016; Pant et al. 2016).

However, for being a successful model, availability of an efficient plant transformation system is often the most limiting step for the majority of the species (Gelvin 2003; Ceasar and Ignacimuthu 2009). Ease of transformation completely contributes the success of Arabidopsis and rice model systems. Therefore, development of novel and efficient transformation approaches and improvement of the existing systems drive the key for the success of *Setaria* as a model (Muthamilarasan and Prasad 2017). Hence, deeper and detailed review of transformation studies in *Setaria* is the need of the hour.

9.3 In Vitro Culture: A Pre-requisite

Tissue culture procedure based on cellular totipotency is not just a theoretical requirement—it is essential for most of the current transformation procedures to achieve a workable efficiency for gene transfer, selection, and regeneration of transformants, although part of the transformation protocol, the regeneration step, is often the biggest hurdle. *Setaria*, in common with other monocots, are also recalcitrant and are difficult targets for regeneration as well as for transformation. Although a few reports are available on *Setaria* tissue culture, the scientific community still encounters numerous problems in obtaining the desired response. The reason is that the morphogenic potential of in vitro culture is determined by several factors, such as age and physiological status of the explants, media composition, carbon source, additional additives in the medium, culture conditions, growth regulators, and the genotype of the donor plants (Ceasar and Ignacimuthu 2008; Sood et al. 2011). Thus, for an efficient use of plant tissue culture techniques, a thorough knowledge of the factors concerned is a pre-requisite.

Plant regeneration from cell cultures proceeds via two paths—somatic embryogenesis and organogenesis. Somatic embryogenesis is the formation of somatic embryos, bipolar structures with the shoot and root meristems, either directly from the explants or a de novo origin from callus. Somatic embryogenesis has become an important part of plant biotechnology research, as the development of transgenic lines in several important crops drives from the somatic embryos. As plants are obtained from single cells, they are mostly normal and lack any phenotypic or genotypic variations. With enhanced understanding of the genetic and physiological factors regulating zygotic and somatic embryogenesis, embryogenic cultures were successfully obtained in a wide variety of species (Braybrook et al. 2006). Somatic embryogenesis is the preferred mode of plant regeneration in the majority of cereals and grasses because of the suitability of the embryogenic calli for *Agrobacterium*-mediated genetic transformation (Kumar et al. 2001), whereas organogenesis is the formation of individual organs as shoots and roots either directly on the explants, or de novo origin from callus. However, comparatively, this is the less preferred mode as cultures retain their morphogenetic potential for only a short time. Plant regeneration through organogenesis is uncommon in millets with reports only available on finger millet and pearl millet (George and Eapen 1990; Jha et al. 2009).

Plant growth regulators play an essential role in regulating the in vitro plant development programmes. Factors such as the (1) type of plant growth regulators, (2) sequence of their application, and (3) exposure timings that determine plant regeneration responses to a large extent vary with the genotype and the species. The ease with which plant material can be manipulated under in vitro conditions opens wide the path for the development of different transformation techniques followed by the recovery of transgenic plants. Another factor encountered during tissue culture is the genotypic variations in culture response. Different genotypes of the same species may respond differently, and conditions favourable for one genotype may not be suitable for another (Bhaskaran and Smith 1990). Hence, selecting genotypes with high regeneration potential is a great obstacle for biotechnological improvement (Plaza-Wuthrich and Tadele 2012).

Furthermore, the type of explants used is also important as it must be suitable for regeneration. In vitro, culture responses vary with different explants from a genotype because of differences in their endogenous hormones level (Pal et al. 2012). A variety of explants such as immature embryos, embryonic cultures, mature seed-derived calli, meristems, excised leaf blades, cotyledons, shoot apices, stem segments, roots, and callus suspension cultures were used as targets in *Setaria* for initiating regenerable cultures. The success of the process to a large extent depends on the age and physiological status of the explant (Birch 1997). However, explants with immature meristematic cells exhibit more tendencies to develop a callus. The selected explants must also have large numbers of regenerable cells with longer regeneration capacity.

The regeneration efficiency is also influenced by the basic media composition as well as by the additional supplements (such as vitamins, amino acids, inorganic nutrients) added to the media. The positive influence of casein hydrolysate, proline, and glutamine on embryogenesis and shoot regeneration was determined by a number of workers in millets and cereals (Wang et al. 2011; Satish et al. 2016). Increasing $CuSO_4$ to a threshold level (in some cases up to $10\times$) resulted in enhanced embryogenic callusing and plant regeneration in various millets (Kothari-Chajer et al. 2008; Sharma et al. 2011). Copper sulfate, being involved with several metabolic activities, photosynthesis, and enzymes might be responsible for the response (Kothari-Chajer et al. 2008). Additionally, $AgNO_3$, a potent ethylene inhibitor, is also successfully employed in various studies, thereby improving callus induction, somatic embryogenesis, and regeneration as well as preventing necrosis (O'Kennedy et al. 2004a).

Optimization and simplification of plant tissue culture methods are therefore often required for enhancing efficiency and reducing culture time. Cells competent for regeneration as well as for transformation are the successful targets for raising transgenic plants (Potrykus 1991). The few competent cells that receive the transgene should quickly recover from the transformation shock and must proliferate and regenerate into complete plants (Sood et al. 2011). However, the recovery of fertile transgenic plants is extremely difficult. Therefore, for quite a long time, optimization of factors governing plant regeneration was the main focus of the monocot transformation researchers (Aulinger et al. 2003). Furthermore, optimized factors and phytohormone treatments successfully increased the competency of recalcitrant explants toward transformation (Geier and Sangwan 1996). Following this, many recalcitrant species were later turned into a pool of successfully transformed plants after identification of potential explants with many regenerable cells and parameter optimization for gene transfer to those cells, together with tailoring selection and regeneration procedures to recover transgenic plants. Transformation efficiency thus directly depends on the tissue culture and gene transfer efficiency (Li et al. 1996).

However, compared to other millets and gramineous crops (namely barley, rice, maize, sorghum, and wheat), there are very few successful reports on regeneration in *Setaria* (Wang et al. 2011).

9.3.1 Cellular Totipotency of *Setaria*

The first ever report on an in vitro study of foxtail millet using anthers as explants was Ban et al. (1971). Later, regeneration of plantlets using immature inflorescences either directly or through somatic embryogenesis was also reported by several researchers (Xu et al. 1983, 1984; Yang and Xu 1985; Vishnoi and Kothari 1996;

Wang et al. 2011). Immature inflorescences are considered to be an important source of totipotent cultures in *S. italica* (Xu et al. 1984; Yang and Xu 1985). However, the use of immature embryos or inflorescences as explants for in vitro studies is constrained because of the need for continuous growth of donor plants. Therefore, regeneration competence of other explants such as shoot apices (Osuna-Avila et al. 1995), leaf base and mesocotyl (Rout et al. 1998), mature seeds (Reddy and Vaidyanath 1990), and immature glumes (Reddy and Vaidyanath 1988, 1990), or indirect regeneration through callus using mature seeds (Rao et al. 1988; Satish et al. 2016) was also explored by several other researchers. Shoot meristem-based regeneration systems are generally genotype-independent with high regeneration potential that can be extended to different genotypes (Osuna-Avila et al. 1995; Rashid 2003; Kishore et al. 2006). On the other hand, mature embryos and seeds are preferred because of their easy handling, storage, and year-round availability. Additionally, the calli induced from the scutellar tissue of mature seeds are competent for regeneration as well as for transformation. Plant regeneration systems were also established for wild relatives of foxtail millet, namely *Setaria lutescens* and *Setaria glauca* (Xu et al. 1983; Diao et al. 1997). However, as per the reports available in the literature, regeneration in *Setaria* is genotype and explant-dependent that is driven by PGRs.

However, for *S. viridis* there was a report by Brutnell et al. (2010) where they induced callus from mature seeds following the protocol published for *S. italica* and *Brachypodium distachyon* (Rao et al. 1988; Rout et al. 1998; Vogel and Hill 2008). Later, Van Eck and Swartwood (2015) claimed to have improved further the Brutnell et al. (2010) protocol by making changes, thereby affecting callus quality and further enhancing transformation efficiency. The in vitro regeneration reports on *Setaria* are summarized in Table 9.1.

9.4 DNA Delivery Systems in *Setaria*

Plant transformation is a targeted gene-based tool that opened new possibilities to modify crops. It has been used extensively for the study of plant functional genomics, namely gene identification, linking genes to biological functions, and investigation of genetically-controlled characteristics. Gene transformation facilitates the introduction of foreign genes into crop plants, diligently developing new genetically-modified organisms and contributing to an overall increase in crop productivity (Sinclair et al. 2004). Furthermore, transgenic technology has emerged as an essential resource for basic scientific research.

Gene transfer can be carried out through either direct or indirect delivery systems. Direct methods are based on physical or chemical processes to deliver naked DNA into the plant cell and are classified as physical (electroporation, particle bombardment, microinjection, liposome fusion, silicon carbide fibers), chemical (PEG-mediated, DEAE dextran-mediated, calcium phosphate precipitation), or DNA imbibition by cells/tissues/organs (Newell 2000; Patnaik and Khurana 2001). Of this microprojectile bombardment, transfer into protoplasts and microinjection are commonly used for plant transformation. Direct delivery methods gained importance as being suitable for both transient and stable gene expression studies, even though the frequency of stable transformation is low. Most of the direct delivery methods were developed and emerged as a key for the successful transformation of monocots or some legumes that were considered to be outside the host range of *Agrobacterium*. Among these methods, biolistics/microprojectile bombardment developed as the most widely accepted, genotype-independent plant transformation system that can be used for diverse species, sub-cellular organelles, cells, protoplasts, pollens, embryos, callus, organized tissues such as meristems, bacteria,

Table 9.1 In vitro regeneration studies reported in *Setaria*

Plant species	Explant used	Media used		Response	Reference
Setaria italica	Anther	Callus induction and sub-culture	2,4-D (1 mg/L) + Kn (1–2 mg/L)	Pollen derived callus and Plant regeneration	Ban et al. (1971)
		Plant regeneration	IAA (2 mg/L) + Kn (2–4 mg/L)		
		Rooting	IAA (2 mg/L) + Kn (0.002–0.1 mg/L)		
Setaria italica; *S. lutescens*	Young spike	Callusing	BA (2 mg/L) or BA (2 mg/L) + IAA (0.2 mg/L) or 2,4-D (2 mg/L) + Kn (0.2 mg/L)	In vitro regeneration	Xu et al. (1983)
Setaria italica	Young inflorescence	Callus induction	2,4-D (2 mg/L) + BA/Kn (0.2–0.5 mg/L)	Plant regeneration through indirect somatic embryogenesis	Xu et al. (1984)
		Callus sub-culture	2,4-D (2 mg/L) + BA/Kn (0.2–0.5 mg/L) and NAA (2 mg/L) + 2iPA (0.2 mg/L) alternate sub-culture		
		Plant regeneration	BA (2 mg/L) + NAA (0.5 mg/L)		
Setaria italica	Suspension culture	Cell suspension	2,4-D (2 mg/L) + Coconut milk (5%)	Somatic embryogenesis and plant regeneration	Yang and Xu (1985)
Setaria italica cv. 315; 212	Mature seeds	Callus induction	2,4-D (2 mg/L) + (0.5 mg/L) Kn	Callusing and plant regeneration; Cv. 315 more competent for regeneration	Rao et al. (1988)
Setaria italica	Mature seeds and immature glumes	Callus induction	LS + 2,4-D (2 mg/L)	Callus and plant regeneration	Reddy and Vaidyanath (1990)
		Plant regeneration	Kn (4 mg/L)		
Setaria italica cv. Nese 2A	Shoot apices	Callus induction	2,4-D (2 mg/L)	Callus and plant regeneration	Osuna-Avila et al. (1995)
		Plant regeneration	2,4-D (0.02 mg/L) + Kn (1 mg/L) + Casein hydrolysate (0.2%)		
Setaria italica	Leaf base and mesocotyl derived from 10-day-old seedling	Callus induction	2,4-D (3 mg/L) + Kn (0.5 mg/L)	Nickel tolerant cell lines developed; plant regeneration via indirect somatic embryogenesis	Rout et al (1998)
		Somatic embryogenesis and plant regeneration	BA (1 mg/L) + Kn (1 mg/L) + 2,4-D (0.5 mg/L)		

(continued)

Table 9.1 (continued)

Plant species	Explant used	Media used		Response	Reference
Setaria italica cv. Jigu 11	Immature inflorescences	Callus induction	2,4-D (2 mg/L) + L-Proline (1 mg/L) + Casein acids hydrolysate (800 mg/L) + $AgNO_3$ (5 mg/L)	Improved plant regeneration for subsequent transformation	Wang et al. (2011)
		Plant regeneration	BA (2 mg/L) + NAA (0.5 mg/L) + L-Proline (1 mg/L) + Casein acids hydrolysate (800 mg/L)		
		Rooting	½ MS + L-Proline (1 mg/L) + Casein acids hydrolysate (800 mg/L)		
Setaria italica genotypes 'CO5', 'CO7', 'TNAU43' and 'RS118'	Mature seeds	Callus induction	2,4-D (3.5 mg/L)	Addition of amino acids, sucrose, maltose and cefotaxime increased the frequency of somatic embryo induction, maturation and plant regeneration in all the four genotypes	Satish et al. (2016)
		Callus sub-culture	2,4-D (3.5 mg/L) + Kn (1 mg/L) + NAA (1 mg/L)		
		Somatic embryogenesis	2,4-D (3.5 mg/L) + Kn (1 mg/L) + NAA (1 mg/L) + Pro (750 mg/L) + Gly (2.0 mg/L) + Arg (150 mg/L) + CEH (800 mg/L)		
		Somatic embryos maturation and plant regeneration	BA (3 mg/L) + 2,4-D (0.2 mg/L), 750 mg/L Pro + Gly (2.0 mg/L) + Arg (150 mg/L) + CEH (800 mg/L)		
		Rooting	½ MS		

fungi, and even animal cells (Altpeter et al. 2005). They allow the transfer of multiple transgenes and large DNA fragments in the target tissue, though DNA integrity is a concern (Barampuram and Zhang 2011). Furthermore, easy handling and regeneration of multiple transformants from one shot favors the method. Moreover, different factors varying from genotype, physiological age, explant, pre- and post-culture period and treatments, and the composition of the culture medium are key regulators driving transformation efficiency. In addition, parameters such as bombardment chamber vacuum pressure, acceleration pressure, the number of bombardments, density and size of microparticles, DNA-micro-particle preparation, and distances such as macro-carrier flight and target, also affect biolistic-mediated transformation efficiency.

Although the biolistic-mediated approach has been successfully used for producing a number of transgenic crops, thereby giving significant impact to agricultural biotechnology, it is definitely not a panacea. Several distinctive factors such as transgene silencing caused by the presence of multiple copies of introduced genes, inability to regenerate plants after bombardment, inefficiency in yielding stable integration events, labour intensive recovery of large numbers of independent transformation events, and high cost limit the utility of this method. In addition, the recalcitrance of many plant species, especially monocots, for efficient regeneration from protoplasts, low transient expression of transgenes, and prolonged tissue culture phases are other disadvantages associated with direct gene delivery methods that employ protoplasts.

On the other hand, some indirect methods are vector-based that employ *Agrobacterium tumefaciens* or *Agrobacterium rhizogenes* for gene transfer into the target cells. *Agrobacterium*-mediated approach is a preferred mode because of stable and simple integration of single or few copies of transgene with little or no rearrangement (Dai et al. 2001; Travella et al. 2005), transfer of larger DNA segments into recipient cells (Hamilton et al. 1996), high transformation efficiency (Ishida et al. 1996; Zhao et al. 1998), and avoiding mosaicism/chimaeras that are common when direct methods transform intact organs. This method is, thus, not associated with problems of transgene instability, gene silencing, and/or co-suppression (Hansen et al. 1997). Advances in transformation methods have extended the use of *Agrobacterium*-mediated approaches for several important monocotyledonous crops, including major cereals and legumes, although the efficiency of gene transfer is still unsatisfactory (Nadolska-Orczyk et al. 2000; Koichi et al. 2002). Cells competent for transformation are often not competent for regeneration and vice versa. *Agrobacterium*-mediated transformation is highly cultivar–dependent, thus limiting to a narrow range of genotypes within a species (Nam et al. 1997).

However, research on transgenesis in millet, especially for *Setaria*, is very limited when compared to other staple cereals. Among the different reports available for millet transformation, biolistics and *Agrobacterium*-mediated methods are the most explored. Optimization of various factors ranging from genotype to explant, *Agrobacterium* strain, and binary vector, a selectable marker, promoter, co-cultivation conditions, and media could be helpful in millets transformation (Dosad and Chawla 2016). The *Agrobacterium* transformation system, being highly cultivar-specific, depends on the screening of most responsive genotypes for tissue culture that is also important for subsequent transformation. Selection of elite genotypes of millets would, thus, enhance their transformation efficiency. Furthermore, the use of the appropriate *Agrobacterium* strain that varies with plant species and cultivar also influences the transformation efficiency. Most of the millet transformations reported the use of *Agrobacterium* strains EHA101, LBA4404, and derivatives of EHA101 (namely EHA105, AGL0, and AGL1) (Plaza-Wuthrich and Tadele 2012). LBA4404 is comparatively superior to EHA105 for the transformation of immature inflorescence-derived calli of millets (Wang et al. 2011). Among promoters, CaMV*35S*, actin, and ubiquitin are commonly used for driving gene expression in various millets such as finger millet, pearl millet, and foxtail millet (Wang et al. 2011; Saha and Blumwald 2016). The use of tissue-specific and inducible promoters is also successfully investigated. Furthermore, the choice of proper selectable marker genes is also important. The most common selectable markers are hygromycin phosphotransferase (*hptII*), neomycin phosphotransferase (*nptII*), and bar gene for millet transformation studies (Girgi et al. 2002; Ceasar and Ignacimuthu 2011; Sharma et al. 2011). A few recent studies also reported the use of phosphomannose isomerase (*manA*) and modified alpha-tubulin gene in pearl millet and finger millet, respectively (O'Kennedy et al. 2004b; Yemets et al. 2008). Use of antioxidants such as DTT and L-cysteine in the co-cultivation media and co-cultivation temperature at a range of 22–28 °C is shown to influence transformation efficiency of foxtail millet transformation (Wang et al. 2003; Liu et al. 2005, 2007). Therefore, optimization of various factors is crucial for developing high-throughput *Setaria* transformation systems.

9.4.1 Biolistic-Mediated Transformation of *Setaria*

Transformation studies reported in foxtail millet use both biolistic- (Dong and Duan 1999, 2000; Liu et al. 2009) and *Agrobacterium tumefaciens*-

mediated approaches (Liu et al. 2005, 2007; Wang et al. 2011). Although the biolistic approach is extensively exploited for cereals, it still lags far behind for *Setaria* with only very few reports available. Diao et al. (1999) reported a detailed study of various factors affecting microprojectile bombardment-mediated transformation of immature inflorescence-derived calli. Later, foxtail millet transgenic plants were developed by bombardment of pollen and inflorescence, although the protocol exhibits very low transformation efficiency (Dong and Duan 1999, 2000). Another study reported the use of plasmids, namely pROKf40s, pROKf40an, and pROKf40i for microprojectile bombardment of *SiPf40* in floret-derived embryogenic calli of foxtail millet (Liu et al. 2009). However, for *S. viridis* no report is available on transgenic development using any of the direct delivery methods.

9.4.2 *Agrobacterium tumefaciens*-Mediated Transformation of *Setaria*

The first report on *Agrobacterium*-mediated genetic transformation of foxtail millet was made by Liu et al. (2005). The kanamycin selected putative transformants were then screened by *gus* gene expression followed by confirmation with Southern hybridization. The authors report transformation efficiency of 6.6% for the protocol. Later, in 2007 Liu et al. reported an ample study investigating several different parameters, namely genotype, explant source, inoculation time, and co-cultivation duration affecting *Agrobacterium*-mediated transformation of foxtail millet. Qin et al. (2008) developed foxtail millet *Si401* transgenic lines through *Agrobacterium*-mediated transformation of panicle-derived calli following the protocol of Liu et al. (2005). The *Si401*, a pollen specific gene, was cloned in pBIN19 under the control of pollen-specific promoter of maize (*Zm13*) and transformed to LBA4404. Silenced *Si401* foxtail millet transgenic lines showed multiple abnormalities during anther development, such as premature degeneration of tapetum, pre-deposition of fibrous bands in endothelium cells, followed by aborted pollen grains. The study demonstrated the role of *Si401* in developing male-sterile plants, breeding foxtail millet hybrid varieties, and uncovering the molecular pathway of cereal anther development. Wang et al. (2011) further optimized the *Agrobacterium*-mediated transformation system of foxtail millet reported by the same group in 2003. The study reported a more detailed, efficient, and reproducible callus induction and plant regeneration system from immature inflorescences. Several factors affecting transformation efficiency such as *Agrobacterium* strain, callus age, co-cultivation temperature, and media composition were investigated in detail. The optimized method was then successfully used to transform foxtail millet with an efficiency of 5.5% with *SBgLR* gene from potato which encodes a lysine-rich protein. Lately, Wang et al. (2014) over-expressed *SiLEA14*, a homolog of the late embryogenesis abundant (LEA) proteins in foxtail millet following the protocol of Qin et al. (2008) and Wang et al. (2011). *SiLEA14* transgenic lines showed improved growth and enhanced salt/drought tolerance. Furthermore, Li et al. (2014, 2017) functionally validated the *SiARDP* and *SiASR4* genes, respectively, in foxtail millet using the protocol of Wang et al. (2011). Together, these studies reveal that both *SiARDP* and *SiASR4* play a critical role in plant adaption to abiotic stresses via an ABA-dependent pathway.

The first preliminary report on the genetic transformation of *S. viridis* by the *Agrobacterium tumefaciens*-mediated method came only recently by Brutnell et al. (2010). They established a transformation protocol for both transient and stable transgene expression in accession A10 of *S. viridis*. Transgenic lines (T0) harboring *gus* gene were developed from seed-derived calli. Transient expression of plastid-localized YFP fusion protein was shown in leaves inoculated with AGL1 strain harboring pPTN469 vector. Similarly, another study by

Van Eck and Swartwood (2015) reported *Agrobacterium*-mediated transformation of calli derived from seeds with a transformation efficiency of about 5%.

9.4.3 Alternative Genetic Transformation Methods

Although biolistic and *Agrobacterium*-mediated DNA delivery methods have been in routine use for plant transformation, a large number of crops still remain unexplored. Furthermore, these methods are laborious and require long tissue culture phases that lead to somaclonal variations and morphological deformities (Lin et al. 2009; Bairu et al. 2011). Also, most of the plant species are highly recalcitrant to in vitro regeneration (Benson 2000). The subsequent technical challenge is thus to reduce time and labor for plant engineering. The ideal solution would be to develop alternative methods that minimize or eliminates the in vitro culture steps and generate high-frequency transformants with predictable and precise transgene expression without collateral genetic damage (Birch 1997). Among alternative methods, in planta transformation holds extensive significance as it reduces labor cost, is less time consuming, and eliminates or reduces the tissue culture phase, is thereby free of somaclonal variations and follows a simplified protocol. This is termed in planta transformation as transgenes are usually transferred into intact plants as naked DNA or through *Agrobacterium*. In planta transformation protocol involves vacuum infiltration, floral dipping, and spraying. The vacuum infiltration method is now routinely used as it enables the access of *Agrobacterium* to deeper plant cells (Bechtold et al. 1993). The vacuum applied generates negative pressure that decreases the air spaces between the plant cells, whereby increase in pressure transfers infiltration medium supplemented with transformation vector into the plant tissue. This method is now routinely used for enhancing in planta transformation efficiency in various crops (Lin et al. 2009; Bai et al. 2013; Mayavan et al. 2013). The floral dip method simply involves dipping of flowering plants into a medium containing the *Agrobacterium* harboring the desired vector with a transgene (Clough and Bent 1998). Floral spray transformation involves spraying (up to three times a day) bolting plants with infiltration media from a distance of 20–30 cm above the plant (Chung et al. 2000). However, alternative methods such as the application of a DNA-pollen mixture to stigmatic surfaces, directly injecting DNA into axial placenta or floral tillers, and direct imbibition by seeds are also reported, but are still impractical because of low reproducibility (Zhou et al. 1983; Langridge et al. 1992; Trick and Finer 1997; Hansen and Wright 1999).

The feasibility of the in planta floral-dip *Agrobacterium*-mediated genetic transformation method in *S. viridis* was confirmed by Martins et al. (2015) and Saha and Blumwald (2016). Transgenic lines were obtained by vacuum infiltrating the boot stage spikes of 1-month-old *S. viridis* plants for 10 min with bacterial suspension harboring the pANIC 6A (Martins et al. 2015). The method reported a transformation efficiency of 0.6%. Unfortunately, the method lacked reproducibility, required prior optimization, and had low transformation efficiency. Saha and Blumwald (2016) came up with a more efficient and optimized rapid spike dip method that enables the high-throughput transformation of *S. viridis*. Different genotypes of *S. viridis* were transformed with five different reporter gene constructs for stable transformation as well as transient gene expression assay using the optimized protocol. Transgenic lines were developed at an efficiency of 0.8 ± 0.1% after dipping 5-day-old S3 spikes in a suspension of *Agrobacterium* supplemented with silwet L-77 (0.025%) and acetosyringone (200 μM). Stable transgenic lines (T1) were obtained within a timeframe of 8–10 weeks. Furthermore, the transgene expression and inheritance was also monitored over generations. The in planta methods reported obviate the necessity of tissue culture phase, thereby accelerating the pace of translational research in a monocot model plant.

9.5 Conclusion and Future Perspectives

The weedy green foxtail (*S. viridis*) and domesticated foxtail millet (*S. italica*) promise to be very informative model species. These models could be used to unravel studies related to artificial selection, C_4 evolution, comparative grass genomics, abiotic stress tolerance, and biomass production in the Panicoid grasses. Both these species hold their exclusive study advantages. Green foxtail plants are comparatively much smaller and can flower when it is just six inches tall with a seed to seed life cycle of 6–8 weeks, whereas foxtail millet is a proper crop so the plants are relatively taller with a longer generation time of about 4 months. Furthermore, being drought- and cold-tolerant can serve as a vast resource of abiotic stress tolerance genes. In common, *Setaria*'s location in the phylogenetic tree is very important, and, with respect to rice, close enough to allow small conserved regulatory sites to be easily recognized but far enough apart to enable randomization of functionless sequences. Furthermore, they are the closest relatives of switchgrass and the invasive weed guinea grass (*Panicum maximum*) and, being in the sister tribe to Andropogoneae (which includes sorghum, maize, sugarcane, and *Miscanthus*), it can serve as a valuable outgroup. Furthermore, the Paniceae clade contains members from all three subtypes of C_4 photosynthesis. C_4 photosynthesis originated multiple times within grasses with independent evolution in Panicoideae (Sinha and Kellogg 1996), so insight is expected. However, several issues need to be addressed to enable *Setaria* to act as an Arabidopsis for the Panicoid grasses. The one major hurdle yet to overcome, however, is transformation. The ease and efficiency of transformation would definitely determine the utility of these model species.

Although few transformation reports are available, scientists still find it difficult, so there is always a continuous quest to find more effectual and economic transformation methods. The major hindrance to functional validation studies is the speed and efficiency of transformation. Conventional transformation methods through the tissue culture phase require technical expertise and a long time-period, and they are costly and involve somaclonal variations. However, recent reports on the floral dip method in *S. viridis* (Martins et al. 2015; Saha and Blumwald 2016) offer a simple protocol that is rapid and low cost. With the availability of in planta transformation reports that it is comparable to the floral dip of Arabidopsis, it is expected that it would facilitate the high-throughput transformation of *S. viridis*. This would further speed up the genetic and genomic studies of important food–feed–fiber–fuel crops. However, transgene expression in some plant parts needs further optimization.

However, an attempt to optimize a highly efficient transformation method for *S. italica* is still on the way. All recent reports on functional validation of genes in *S. italica* used the protocol of Wang et al. (2011) only. The method gave a transformation efficiency of 5.5%, which is not reasonable for a model plant. Therefore, the focus should be toward making improvements to the widely used existing tissue culture-based transformation methods together with finding the probability for in planta approaches. Parallel research for enhancing regeneration followed by transformation of a wide range of target tissues and genotypes is urgently needed. Factors influencing the in vitro response of target tissues need a deeper understanding. Approaches for simplifying and shortening the tissue culture phase by manipulating cell and tissue development are of interest and should be explored. Furthermore, genotype-independent methods must be developed for widening its applicability. Transformation parameters must be optimized for each biological target employed. Use of plant genes concerned with dedifferentiation, wound response, and/or homologous recombination that improves the recovery of transformed plants should be enhanced (Altpeter et al. 2016). Moreover, understanding factors affecting *Agrobacterium*-mediated transformation would further help in the optimization of transformation. The use of other microbes such as *Rhizobium* spp. and *Ensifer adhaerens* could facilitate the transformation of various crop plants

(Broothaerts et al. 2005; Zuniga-Soto et al. 2015). Several genome-wide studies have recently been executed in *Setaria* crops which have identified interesting candidate genes with roles in molecular stress response (Muthamilarasan et al. 2014a, b, c, 2015a, b, 2016b), and these genes are awaiting over-expressions studies in foxtail millet. Together with efficient and improved traditional transformation approaches, the in planta method could be a boon for the *Setaria* scientific community. Determining factors such as infiltration medium and method, stage of plant growth and development, surfactant concentration, and *Agrobacterium* strains could be important for in planta transformation optimization. These methods would rapidly hasten the genetics and genomics study in *Setaria* and enable translation of the basic understanding of plant biology to crop improvement of Pooid and Panicoid, agriculturally the most important clades of grasses (Brutnell et al. 2015). Therefore, in conclusion, it can be said that these model species hold tremendous potential as an agricultural solution for a world facing climate change with limited natural resources.

Acknowledgements Studies on millet genomics in Dr. Manoj Prasad's laboratory are supported by Science and Engineering Research Board (SERB), Department of Science and Technology (DST), Govt. of India [Grant No. EMR/2015/000464], by Department of Biotechnology, Govt. of India [Grant No. BT/HRD/NBA/37/01/2014], and by Core Grant of National Institute of Plant Genome Research (NIPGR), New Delhi, India. Priyanka Sood acknowledges the Young Scientist Award from DST-SERB, Govt. of India [File No. YSS/2014/000870/LS].

References

Altpeter F, Varshney A, Abderhalden O, Douchkov D, Sautter C, Kumlehn J, Dudler R, Schweizer P (2005) Stable expression of a defense-related gene in wheat epidermis under transcriptional control of a novel promoter confers pathogen resistance. Plant Mol Biol 57:271–283

Altpeter F, Springer NM, Bartley LE, Blechl AE, Brutnell TP, Citovsky V, Conrad LJ, Gelvin SB, Jackson DP, Kausch AP et al (2016) Advancing crop transformation in the era of genome editing. Plant Cell 28:1510–1520. doi:10.1105/tpc.16.00196

Amadou I, Gounga ME, Le GW (2013) Millets: nutritional composition, some health benefits and processing-A review. Emirates J Food Agric 25:501–508. doi:10.9755/ejfa.v25i7.12045

Aulinger I, Peter S, Schmid J, Stamp P (2003) Gametic embryos of maize as a target for biolistic transformation: comparison to immature zygotic embryos. Plant Cell Rep 21:585–591. doi:10.1007/s00299-002-0556-7

Bai H, Cao Y, Quan J, Dong L, Li Z, Zhu Y, Zhu L, Dong Z, Li D (2013) Identifying the genome-wide sequence variations and developing new molecular markers for genetics research by re-sequencing a landrace cultivar of foxtail millet. PLoS ONE 8: e73514. doi:10.1371/journal.pone.0073514

Bairu MW, Aremu AO, VanStaden J (2011) Somaclonal variation in plants: causes and detection methods. Plant Growth Regul 63:147–173

Ban Y, Kokuba T, Miyaji Y (1971) Production of haploid plant by anther culture of *Setaria italica*. Bull Fac Agric Kagoshima Univ 21:77–81

Barampuram S, Zhang ZJ (2011) Recent advances in plant transformation. Methods Mol Biol 701:1–35

Bechtold N, Ellis J, Pelletier G (1993) *In planta Agrobacterium*-mediated gene transfer by infiltration of adult *Arabidopsis thaliana* plants. C R Acad Sci Paris Life Sci 316:1194–1199

Benabdelmouna A, Shi Y, Abirached-Darmency M, Darmency H (2001) Genomic in situ hybridization (GISH) discriminates between the A and the B genomes in diploid and tetraploid *Setaria* species. Genome 44:685–690. doi:10.1139/gen-44-4-685

Bennetzen JL, Schmutz J, Wang H, Percifield R, Hawkins J, Pontaroli AC, Estep M, Feng L, Vaughn JN et al (2012) Reference genome sequence of the model plant *Setaria*. Nat Biotechnol 30:555–561. doi:10.1038/nbt.2196

Benson EE (2000) *In vitro* plant recalcitrance: an introduction. Vitro Cell Dev Biol Plant 36(3):141–148

Bhaskaran S, Smith RH (1990) Regeneration in cereal tissue culture: a review. Crop Sci 30:1328–1336

Birch RG (1997) Plant transformation: problems and strategies for practical application. Annu Rev Plant Physiol Plant Mol Biol 48:297–326

Braybrook SA, Stone SL, Park S, Bui AQ, Le BH, Fischer RL, Goldberg RB, Harada JJ (2006) Genes directly regulated by *LEAFY COTYLEDON2* provide insight into the control of embryo maturation and somatic embryogenesis. Proc Natl Acad Sci USA 103:3468–3473

Broothaerts W, Mitchell HJ, Weir B, Kaines S, Smith LMA, Yang W, Mayer JE, Roa-Rodríguez C, Jefferson RA (2005) Gene transfer to plants by diverse species of bacteria. Nature 433:629–633

Brutnell TP (2015) Model grasses hold key to crop improvement. Nat Plants 1:15062. doi:10.1038/nplants.2015.62

Brutnell TP, Wang L, Swartwood K, Goldschmidt A, Jackson D, Zhu XG, Kellogg E, Van Eck J (2010) *Setaria viridis*: a model for C_4 photosynthesis. Plant Cell 22:2537–2544

Brutnell TP, Bennetzen JL, Vogel JP (2015) *Brachypodium distachyon* and *Setaria viridis*: model genetic systems for the grasses. Annu Rev Plant Biol 66:465–485. doi:10.1146/annurev-arplant-042811-105528

Ceasar SA, Ignacimuthu S (2008) Efficient somatic embryogenesis and plant regeneration from shoot apex explants of different Indian genotypes of finger millet (*Eleusine coracana* (L.) Gaertn.). In Vitro Cell Dev Biol Plant 44:427–435

Ceasar SA, Ignacimuthu S (2009) Genetic engineering of millets: current status and future prospects. Biotechnol Lett 31:779–788

Ceasar SA, Ignacimuthu S (2011) *Agrobacterium*-mediated transformation of finger millet (*Eleusine coracana* (L.) Gaertn.) using shoot apex explants. Plant Cell Rep 30:1759–1770. doi:10.1007/s00299-011-1084-0

Ceccarelli S, Grando S (1996) Drought as a challenge for the plant breeder. Plant Growth Regul 20:149–155. doi:10.1007/BF00024011

Changmei S, Dorothy J (2014) Millet-the frugal grain. Int J Sci Res Rev 3:75–90

Chung MH, Chen MK, Pan SM (2000) Floral spray transformation can efficiently generate *Arabidopsis* transgenic plants. Transgenic Res 9:471–476

Clough SJ, Bent AF (1998) Floral dip: a simplified method for *Agrobacterium*-mediated transformation of *Arabidopsis thaliana*. Plant J 16(6):735–743

Dai A (2011) Drought under global warming: a review. Wiley Interdiscip Rev Clim Change 2:45–65

Dai S, Zheng P, Marmey P, Zhang S, Tian W, Chen S, Beachy RN, Fauquet C (2001) Comparative analysis of transgenic rice plants obtained by *Agrobacterium*-mediated transformation and particle bombardment. Mol Breed 7:25–33

Defelice MS (2002) Green foxtail, *Setaria viridis* (L.) P. Beauv. Weed Technol 16:253–257

Diao X, Duan S, Zhao L, Chen Z (1997) Tissue culture and plantlet regeneration of *Setaria glauca*. Plant Cell Rep 33(2):128–129

Diao X, Chen Z, Duan S, Liu Y, Zhao L, Sun J (1999) Factors influencing foxtail millet embryogenic calli transformation by particle bombardment. Acta Agric Boreali Sin 14(3):31–36

Diao X, Schnable J, Bennetzen JL, Li J (2014) Initiation of *Setaria* as a model plant. Front Agric Sci Eng 1:16. doi:10.15302/J-FASE-2014011

Dong Y, Duan S (1999) Establishment of embryogenic cell suspension culture and plant regeneration of millet and gene transfer. J Basic Sci Eng 7(1):34–40

Dong Y, Duan S (2000) Production of transgenic millet plants via particle bombardment. Acta Bot Boreal-Occident Sin 20(2):175–178

Dosad S, Chawla HS (2016) *In vitro* plant regeneration and transformation studies in millets: current status and future prospects. Indian J Plant Physiol 21 (3):239–254

Doust AN, Kellogg EA, Devos KM, Bennetzen JL (2009) Foxtail millet: a sequence-driven grass model system. Plant Physiol 149:137–141

Dwivedi S, Upadhyaya H, Senthilvel S, Hash C, Fukunaga K, Diao X, Santra D, Baltensperger D, Prasad M (2012) Millets: genetic and genomic resources. In: Janick J (ed) Plant breeding reviews. Wiley-Blackwell, New York, pp 247–375

Fang X, Dong K, Wang X, Liu T, He J, Ren R, Zhang L, Liu R, Liu X, Li M, Huang M, Zhang Z, Yang T (2016) A high density genetic map and QTL for agronomic and yield traits in foxtail millet [*Setaria italica* (L.) P. Beauv.]. BMC Genom 17:336

Geier T, Sangwan RS (1996) Histology and chimeral segregation reveal cell-specific differences in the competence for shoot regeneration and *Agrobacterium*-mediated transformation in *Kohleria* internode explants. Plant Cell Rep 15:386–390

Gelvin SB (2003) *Agrobacterium*-mediated plant transformation: the biology behind the 'gene-jockeying' tool. Microbiol Mol Biol Rev 67:16–37

George L, Eapen S (1990) High frequency plant-regeneration through direct shoot development and somatic embryogenesis from immature inflorescence cultures of finger millet (*Eleusine coracana* Gaertn). Euphytica 48:269–274

Girgi M, O'Kennedy MM, Morgenstern A, Smith G, Lörz H, Oldach KH (2002) Transgenic and herbicide resistant pearl millet (*Pennisetum glaucum* L.) R. Br. via microprojectile bombardment of scutellar tissue. Mol Breed 10:243–252

Goron TL, Raizada MN (2015) Genetic diversity and genomic resources available for the small millet crops to accelerate a new green revolution. Front Plant Sci 6:157

Hamilton CM, Frary A, Lewis C, Tanksley SD (1996) Stable transfer of intact high molecular weight DNA into plant chromosomes. Proc Natl Acad Sci USA 93:9975–9979. doi:10.1073/pnas.93.18.9975

Hansen G, Wright MS (1999) Recent advances in the transformation of plants. Trends Plant Sci 4:226–230. doi:10.1016/S1360-1385(99)01412-0

Hansen G, Shillito RD, Chilton MD (1997) T-strand integration in maize protoplasts after co-delivery of a T-DNA substrate and virulence genes. Proc Natl Acad Sci USA 94:11726–11730

Hodge JG, Kellogg EA (2016) Abscission zone development in *Setaria viridis* and its domesticated relative, *Setaria italica*. Amer J Bot 103:998–1005

Huang P, Feldman M, Schroder S, Bahri BA, Diao X, Zhi H, Estep M, Baxter I, Devos KM, Kellogg EA (2014) Population genetics of *Setaria viridis*, a new model system. Mol Ecol 23:4912–4925. doi:10.1111/mec.12907

Ishida Y, Saito H, Ohta S, Hiei Y, Komari T, Kumashiro T (1996) High efficiency transformation of maize (*Zea mays* L.) mediated by *Agrobacterium tumefaciens*. Nat Biotechnol 14:745–750

Jha P, Yadav CB, Anjaiah V, Bhat V (2009) *In vitro* plant regeneration through somatic embryogenesis and direct shoot organogenesis in *Pennisetum glaucum* (L.) R. Br. In Vitro Cell Dev Biol Plant 45:145–154

Jia G, Huang X, Zhi H, Zhao Y, Zhao Q, Li W, Chai Y, Yang L, Liu K, Lu H, Zhu C, Lu Y, Zhou C, Fan D,

Weng Q et al (2013) A haplotype map of genomic variations and genome-wide association studies of agronomic traits in foxtail millet (*Setaria italica*). Nat Genet 45:957–961. doi:10.1038/ng.2673

Kishore SN, Visarada KBRS, Aravinda Lakshmi Y, Pashupatinath E, Rao SV, Seetharama N (2006) In vitro culture methods in Sorghum with shoot tips the explants material. Plant Cell Rep 25:174–182. doi:10.1007/s00299-005-0044-y

Koichi T, Bae CH, Seo MS, Song IJ, Lim YP, Song PS, Lee HY (2002) Overcoming of barriers to transformation in monocot plants. J Plant Biotech 4:135–141

Kothari-Chajer A, Sharma M, Kachhwaha S, Kothari SL (2008) Micronutrient optimization results into highly improved *in vitro* plant regeneration in kodo (*Paspalum scrobiculatum* L.) and finger (*Eleusine coracana* (L.) Gaertn.) millets. Plant Cell Tissue Org Cult 94(6):105–112

Kumar S, Agarwal K, Kothari SL (2001) In vitro induction and enlargement of apical domes and formation of multiple shoots in finger millet, *Eleusine coracana* (L.) Gaertn and crowfoot grass, *Eleusine indica* (L.) Gaertn. Curr Sci 81:1482–1485

Langridge P, Brettschneider R, Lazzeri P, Lorz H (1992) Transformation of cereals via *Agrobacterium* and pollen tube pathway: a critical assessment. Plant J 2:631–638

Li P, Brutnell TP (2011) *Setaria viridis* and *Setaria italica*, model genetic systems for the panicoid grasses. J Exp Bot 62:3031–3037

Li HQ, Sautter C, Potrykus I, Puonti-Kaerlas J (1996) Genetic transformation of cassava (*Manihot esculenta* Crantz). Nat Biotechnol 14:736–740

Li C, Yue J, Wu X, Xu C, Yu J (2014) An ABA-responsive DRE-binding protein gene from *Setaria italica*, *SiARDP*, the target gene of *SiAREB*, plays a critical role under drought stress. J Exp Bot 65:5415–5427. doi:10.1093/jxb/eru302

Li ZY, Wang N, Dong L, Bai H, Quan JZ, Liu L, Dong ZP (2015) Differential gene expression in foxtail millet during incompatible interaction with *Uromyces Setariae-italicae*. PLoS ONE 10:e0123825. doi:10.1371/journal.pone.0123825

Li J, Dong Y, Li C, Pan Y, Yu J (2017) *SiASR4*, the target gene of SiARDP from *Setaria italica*, improves abiotic stress adaption in plants. Front Plant Sci 7:2053

Lin J, Zhou B, Yang Y, Mei J, Zhao X, Guo X, Huang X, Tang D, Liu X (2009) Piercing and vacuum infiltration of the mature embryo: a simplified method for *Agrobacterium*-mediated transformation of indica rice. Plant Cell Rep 28(7):1065–1074

Liu Y, Yu J, Zhao Q, Zhu D, Ao G (2005) Genetic transformation of millet (*Setaria italica*) by *Agrobacterium*-mediated. Chin J Agr Biotechnol 13:32–37

Liu YH, Yu JJ, Ao GM, Zhao Q (2007) Factors influencing *Agrobacterium*-mediated transformation of foxtail millet (*Setaria italica*). Chin J Biochem Mol Biol 23:531–536

Liu Y, Feng X, Xu Y, Yu J, Ao G, Peng Z, Zhao Q (2009) Overexpression of millet ZIP-like gene (*SiPf40*) affects lateral bud outgrowth in tobacco and millet. Plant Physiol Biochem 47:1051–1060

Liu X, Tang S, Jia G, Schnable JC, Su H, Tang C, Zhi H, Diao X (2016) The C-terminal motif of SiAGO1b is required for the regulation of growth, development and stress responses in foxtail millet (*Setaria italica* (L.) P. Beauv). J Exp Bot 67:erw135

Mandadi KK, Pyle JD, Scholthof KBG (2014) Comparative analysis of antiviral responses in *Brachypodium distachyon* and *Setaria viridis* reveal conserved and unique outcomes among C_3 and C_4 plant defenses. Mol Plant-Microbe Interact 27:1277–1290. doi:10.1094/MPMI-05-14-0152-R

Martin AP, Palmer WM, Brown C, Abel C, Lunn JE, Furbank RT, Grof CPL (2016) A developing *Setaria viridis* internode: an experimental system for the study of biomass generation in a C_4 model species. Biotechnol Biofuels 9:45

Martins PK, Nakayama TJ, Ribeiro AP, Cunha BADBd, Nepomuceno AL, Harmon FG, Kobayashi AK, Molinari HBC (2015) *Setaria viridis* floral-dip: a simple and rapid *Agrobacterium*-mediated transformation method. Biotechnol Rep 6:61–63

Mauro-Herrera M, Doust AN (2016) Development and genetic control of plant architecture and biomass in the panicoid grass, Setaria. PLoS ONE 11:e0151346

Mayavan S, Subramanyam K, Arun M, Rajesh M, Dev GK, Sivanandhan G, Jaganath B, Manickavasagam M, Selvaraj N, Ganapathi A (2013) *Agrobacterium tumefaciens*-mediated *in planta* seed transformation strategy in sugarcane. Plant Cell Rep 32(10):1557–1574

Muthamilarasan M, Prasad M (2015) Advances in *Setaria* genomics for genetic improvement of cereals and bioenergy grasses. Theor Appl Genet 128:1–14

Muthamilarasan M, Prasad M (2017) Genetic determinants of drought stress tolerance in *Setaria*. In: Doust A, Diao X (eds) Genetics and genomics of Setaria. Springer, pp 267–289

Muthamilarasan M, Suresh BV, Pandey G, Kumari K, Parida SK, Prasad M (2014a) Development of 5123 intron-length polymorphic markers for large-scale genotyping applications in foxtail millet. DNA Res 21:41–52

Muthamilarasan M, Khandelwal R, Yadav CB, Bonthala VS, Khan Y, Prasad M (2014b) Identification and molecular characterization of MYB transcription factor superfamily in C_4 model plant foxtail millet (*Setaria italica* L.). PLoS ONE 9:e109920

Muthamilarasan M, Bonthala VS, Mishra AK, Khandelwal R, Khan Y, Roy R, Prasad M (2014c) C_2H_2-type of zinc finger transcription factors in foxtail millet define response to abiotic stresses. Funct Integr Genom 14:531–554

Muthamilarasan M, Bonthala VS, Khandelwal R, Jaishakar J, Shweta S, Nawaz K, Prasad M (2015a) Global analysis of WRKY transcription factor superfamily in *Setaria* identifies potential candidates

involved in abiotic stress signaling. Front Plant Sci 6:910
Muthamilarasan M, Khan Y, Jaishankar J, Shweta S, Lata C, Prasad M (2015b) Integrative analysis and expression profiling of secondary cell wall genes in C_4 biofuel model *Setaria italica* reveals targets for lignocellulose bioengineering. Front Plant Sci 6:965
Muthamilarasan M, Dhaka A, Yadav R, Prasad M (2016a) Exploration of millet models for developing nutrient rich graminaceous crops. Plant Sci 242:89–97
Muthamilarasan M, Mangu VR, Zandkarimi H, Prasad M, Baisakh N (2016b) Structure, organization and evolution of ADP-ribosylation factors in rice and foxtail millet, and their expression in rice. Sci Rep 6:24008
Nadolska-Orczyk A, Orczyk W, Przetakiewicz A (2000) *Agrobacterium*-mediated transformation of cereals–from technique development to application. Acta Physiol Plant 22:77–88
Nam J, Matthysse AG, Gelvin SB (1997) Differences in susceptibility of Arabidopsis ecotypes to crown gall disease may result from a deficiency in T-DNA integration. Plant Cell 9:317–333
Newell CA (2000) Plant transformation technology; developments and applications. Mol Biotechnol 16:53–65
O'Kennedy MM, Smith G, Botha FC (2004a) Improved regeneration efficiency of a pearl millet (*Pennisetum glaucum* [L.] R. Br.) breeding line. South Afr J Bot 70 (4):502–508
O'Kennedy MM, Burger JT, Botha FC (2004b) Pearl millet transformation system using the positive selectable marker gene phosphomannose isomerase. Plant Cell Rep 22:684–690
O'Kennedy MM, Grootboom A, Shewry PR (2006) Harnessing sorghum and millet biotechnology for food and health. J Cereal Sci 44:224–235
Osuna-Avila P, Nava-Cedillo A, Jofre-Garfias AE, Cabrera-Ponce JL (1995) Plant regeneration from shoot apex explant of foxtail millet. Plant Cell Tissue Org Cult 40:33–35
Pal AK, Acharya K, Ahuja PS (2012) Endogenous auxin level is a critical determinant for *in vitro* adventitious shoot regeneration in potato (*Solanum tuberosum* L.). J Plant Biochem Biotechnol 21:205–212
Pant SR, Irigoyen S, Doust AN, Scholthof KG, Mandadi KK (2016) *Setaria*: a food crop and translational research model for C_4 grasses. Front Plant Sci 15:1885
Patnaik D, Khurana P (2001) Wheat biotechnology: a minireview. Eur J Biotech 4:1–29
Plaza-Wuthrich S, Tadele Z (2012) Millet improvement through regeneration and transformation. Biotechnol Mol Biol Rev 7(2):48–61
Potrykus I (1991) Gene transfer to plants: assessment of published approaches and results. Annu Rev Plant Physiol Plant Mol Biol 42:205–225
Qin F, Zhao Q, Ao G, Yu J (2008) Co-suppression of *Si401* a maize pollen specific *Zm401* homologous gene, results in aberrant anther development in foxtail millet. Euphytica 163(1):103–111. doi:10.1007/s10681-007-9610-4
Rachie KO (1975) The millet: importance, utilization and outlook. International Crop Research Institute for Arid Tropics, Hyderabad, India
Rao AM, Kavi Kishor PB, Ananda Reddy L, Vaidvanath K (1988) Callus induction and high frequency plant regeneration in Italian millet (*Setaria italica*). Plant Cell Rep 7:557–559
Reddy LA, Vaidyanath K (1988) Regeneration of foxtail millet plants from calli derived from immature glumes. Indian J Plant Physiol 31:290–292
Reddy LA, Vaidyanath K (1990) Callus formation and regeneration in two induced mutants of foxtail millet (*Setaria italica*). J Genet Breed 44:133–138
Rout GR, Samataray S, Das D (1998) In vitro selection and characterization of Ni-tolerant callus lines of *Setaria italica* L. Acta Physiol Plant 20:269–275
Saha P, Blumwald E (2016) Spike-dip transformation of *Setaria viridis*. Plant J 86:89–101
Saha D, Dipnarayan S, Channabyre Gowda MV, Lalit A, Manjusha V, Bansal KC (2016) Genetic and genomic resources of small millets. CRC Crit Rev Plant Sci 35:56–79
Satish L, Rency AS, Rathinapriya P, Ceasar SA, Pandian S, Rameshkumar R, Rao TB, Balachandran SM, Ramesh M (2016) Influence of plant growth regulators and spermidine on somatic embryogenesis and plant regeneration in four Indian genotypes of finger millet (Eleusinecoracana (L.) Gaertn). Plant Cell Tissue Org Cult 124:15–31
Sharma M, Kothari-Chajer A, Jagga-Chugh S, Kothari SL (2011) Factors influencing *Agrobacterium tumefaciens*-mediated genetic transformation of *Eleusine coracana* (L.) Gaertn. Plant Cell Tissue Org Cult 105:93–104
Sinclair TR, Purcell LC, Sneller CH (2004) Crop transformation and the challenge to increase yield potential. Trends Plant Sci 9:70–75
Sinha NR, Kellogg EA (1996) Parallelism and diversity in multiple origins of C_4 photosynthesis in the grass family. Am J Bot 83:1458–1470
Sood P, Bhattacharya A, Sood A (2011) Problems and possibilities of monocot transformation. Biol Plantarum 55:1–15
Tadele Z (2016) Drought adaptation in millets. In: Shanker AK, Shanker C (eds) Abiotic and biotic stress in plants—recent advances and future perspectives. InTech, Rijeka, pp 639–662. doi:10.5772/61929
Travella S, Ross SM, Harden J, Everett C, Snape JW, Harwood WA (2005) A comparison of transgenic barley lines produced by particle bombardment and *Agrobacterium*-mediated techniques. Plant Cell Rep 23:780–789
Trick HN, Finer JJ (1997) SAAT: sonication-assisted *Agrobacterium*-mediated transformation. Transgenic Res 6:329–336
Van Eck J, Swartwood K (2015) *Setaria viridis*. Methods Mol Biol 1223:57–67
Rashid VA (2003) Somatic embryogenesis or shoot formation following high 2,4-D pulse-treatment of

mature embryos of *Paspalum scrobiculatum*. Biol Plant 46:297–300

Vishnoi RK, Kothari SL (1996) Somatic embryogenesis and efficient plant regeneration in immature inflorescence culture of *Setaria italica* (L.) Beauv. Cereal Res Commun 24(3):291–297

Vogel J, Hill T (2008) High-efficiency *Agrobacterium*-mediated transformation of *Brachypodium distachyon* inbred line Bd21–3. Plant Cell Rep 27:471–478

Wang Y, Li W, Diao X (2003) Genetic transformation of foxtail millet mediated by *Agrobacterium tumefaciens*. J Hebei Agric Sci 7(4):1–6

Wang M, Pan Y, Li C, Liu C, Zhao Q, Ao GM, Yu JJ (2011) Culturing of immature inflorescences and *Agrobacterium*-mediated transformation of foxtail millet (*Setaria italica*). Afr J Biotechnol 10:16466–16479. doi:10.5897/ajb10.2330

Wang M, Li P, Li C, Pan Y, Jiang X, Zhu D, Zhao Q, Yu J (2014) SiLEA14, a novel atypical LEA protein, confers abiotic stress resistance in foxtail millet. BMC Plant Biol 14(1):290. doi:10.1186/s12870-014-0290-7

Xu Z, Wei Z, Yang L (1983) Tissue culture of *Setaria italica* and *Setaria lutescens*. Plant Physiol Commun 5:40

Xu Z, Wang D, Yang L, Wei Z (1984) Somatic embryogenesis and plant regeneration in callus cultured immature inflorescence of *Setaria italica*. Plant Cell Rep 3:149–150

Xue CX, Zhi H, Fang X, Liu X, Tang S, Chai Y, Zhao B, Jia G, Diao X (2016) Characterization and fine mapping of *SiDWARF2* (*D2*) in foxtail millet. Crop Sci 56:95–103. doi:10.2135/cropsci2015.05.0331

Yang L, Xu Z (1985) Somatic embryogenesis and plant regeneration from cell suspension culture of *Setaria italica* (L.) Beauv. Acta Biologiae Experimentalis Sin 18(4):493–498

Yang X, Wan Z, Perry L, Lu H, Wang Q, Zhao C, Li J, Xie F, Yu J, Cui T, Wang T, Li M, Ge Q (2012) Early millet use in northern China. Proc Natl Acad Sci USA 109:3726–3730

Yemets A, Sheremet Y, Vissenberg K, Van Orden J, Verbelen JP, Blume YB (2008) Effects of tyrosine kinase and phosphatase inhibitors on microtubules in *Arabidopsis* root cells. Cell Biol Int 32:630–637

Zhang G, Liu X, Quan Z, Cheng S, Xu X, Pan S, Xie M, Zeng P, Yue Z, Wang W, Tao Y, Bian C, Han C et al (2012) Genome sequence of foxtail millet (*Setaria italica*) provides insights into grass evolution and biofuel potential. Nat Biotechnol 30:549–554. doi:10.1038/nbt.2195

Zhao ZY, Gu W, Cai T, Tagliani LA, Hondred DA, Bond D, Krell S, Rudert ML, Bruce WB, Pierce DA (1998) Molecular analysis of T0 plants transformed by *Agrobacterium* and comparison of *Agrobacterium*-mediated transformation with bombardment transformation in maize. Maize Genet Coop Newslett 72:34–37

Zhou GY, Weng J, Zhen YS, Huang JG, Qian SY, Liu GL (1983) Introduction of exogenous DNA into cotton embryos. In: Wu R, Grossman L, Molddave K (eds) Methods in enzymology, recombination DNA, Part C, vol 101. Academic Press, New York, pp 433–481

Zuniga-Soto E, Mullins E, Dedicova B (2015) *Ensifer* mediated transformation: an efficient non-*Agrobacterium* protocol for the genetic modification of rice. Springerplus 4:600. doi:10.1186/s40064-015-1369-9

Nutrition Potential of Foxtail Millet in Comparison to Other Millets and Major Cereals

10

Tirthankar Bandyopadhyay, Vandana Jaiswal and Manoj Prasad

Abstract

Global population is burgeoning at an alarming rate and is expected to reach 9.7 billion by 2050 and 11.2 billion by the end of this century. This has led to immense pressure on global agriculture, compounded by dwindling productivity of the existing systems and acreage because of climate change, resulting in ever-increasing input costs for the cultivation of most resource-intensive cereal crops such as rice, wheat, and maize. Ironically, the most affected populations are those with least resources to mitigate the problem—those belonging to Asian and Sub-Saharan Africa. It is against this backdrop that there is an ever-increasing need for adopting cereal crops that are easy to cultivate, less resource hungry, climate resilient, and importantly, that meet the major nutritional requirement of the feeding population. Foxtail millet is a perfect cereal crop in this light and stands to help significantly global endeavors toward food security and nutrition. The present chapter provides a comparative nutritional assessment of foxtail millet with other cereal crops, summarizes the major scientific approaches currently being undertaken for its biofortification and highlights potential avenues of crop improvement using conventional breeding, genomics, and other interdisciplinary "omic" tools.

10.1 Introduction

Food and nutrition security has emerged as one of the most crucial challenges confronting global agriculture over the last two decades, compounded by the burgeoning world population. Plants are the major providers of nutrients for a majority of this population, and are indispensable for normal growth and development. Ironically,

T. Bandyopadhyay (✉) · M. Prasad (✉)
National Institute of Plant Genome Research (NIPGR), Aruna Asaf Ali Marg, New Delhi 110067, India
e-mail: manoj_prasad@nipgr.ac.in

V. Jaiswal (✉)
Laboratory of Translational and Evolutionary Genomics, School of Life Science, Jawaharlal Nehru University, New Delhi 110067, India

M. Prasad (ed.), *The Foxtail Millet Genome*, Compendium of Plant Genomes,
DOI 10.1007/978-3-319-65617-5_10

with the growing population and climate change, input costs for growing conventional resource-intensive cereal crops such as rice, wheat, and maize have risen, rendering them unreachable to many poor feeding populations in Asia and Sub-Saharan Africa. Furthermore, such crops are unable to meet major nutritional requirements for the great majority of the population (Deaton and Dreze 2009) and are susceptible to key biotic and abiotic stresses (Lata et al. 2013). Nearly half of the world's population, mainly from sub-Saharan Africa and Asia, suffer from malnutrition because of their reliance on conventional cereal crops (White and Broadley 2005; Hirschi 2009; Zhao and McGrath 2009). Stunting and below-average weight are two common symptoms of malnutrition, which affect children more than adults. Worldwide, 161 million adults and 99 million children below the age of 5 years are projected to have stunted growth and below-average weight, respectively. Furthermore, almost half of such stunted and two-third of below-average-weight children live in Asia and one-third of such children live in Africa (http://data.unicef.org/nutrition/malnutrition). Unfortunately, about half of the mortality in such children is attributed to malnutrition worldwide (Black et al. 2013). India accounts for the highest population of malnourished children (Jayachandran and Pande 2013).

Millets, commonly known as "small seeded grasses", are an agronomically important C4 cereal crop within the grass subfamily panicoideae, which include maize, sorghum, finger millet, kodo millet, proso millet, barnyard millet, and little millet. They are widely cultivated in sub-Saharan Africa and Asia as well as in Argentina and USA (Dwivedi et al. 2011). Among them, foxtail millet [*Setaria italica* (L)] is the second largest crop after pearl millet [*Eleusine coracana* (L.) Gaertn] cultivated for food in Asia and as forage crops in Europe, North America, Australia, and North Africa (Austin 2006; Muthamilarasan and Prasad 2015). They are known to be more nutritive than rice or wheat as their seeds contain a relatively higher amount of proteins, essential amino acids, major nutritionally important elements such as iron, zinc, phosphorus, potassium, calcium, and vitamin B (Hegde et al. 2005; Saleh et al. 2013). Furthermore, millets are better adapted to climate change as they can thrive efficiently under minimal conditions of soil fertility, moisture, and higher temperature (Lata et al. 2013; Bergamini et al. 2013)

In addition to requiring higher resource inputs for their cultivation and being nutritionally poorer, most conventional cereal grains (such as rice and wheat) have higher glycemic indices (GI) (leading to hyperglycemia upon consumption) than millets. WHO estimates that 422 million people worldwide have diabetes with an estimated 3.7 million deaths caused directly or indirectly by high blood glucose in 2012. Interestingly, a majority of such cases were accounted for by middle- and low-income countries (http://www.who.int/mediacentre/factsheets/fs312/en/). Despite many innovative and useful measures that have been taken to contain this menace (such as supplementary diets, immunization, and lifestyle changes), a comprehensive solution to the problem is yet to be realized. The introduction of low-GI, nutritious, cheap alternative stable food appears to be the most viable and potent approach to address the problem, wherein millets stand to play a big role.

In view of the above, the present chapter highlights the most salient features of foxtail millet of use in addressing the pressing issues of food security, climate change, human nutrition, and stress resilience confronting today's agriculture. The chapter discusses the nutritional competence of the crop for human and forage requirements as well as its potential role in facilitating crop improvement efforts through conventional breeding and transgenic approaches.

10.2 Nutritional Profile of Foxtail Millet

Nutritionally, foxtail millet has a higher nutritive worth than major cereals such as wheat and rice (Parameswaran and Sadasivam 1994). It has a higher content of essential amino acids (apart from lysine and methionine), phytochemicals and micronutrients (Mal et al. 2010; Singh and Raghuvanshi 2012) and antioxidants compared to non-millet cereals and are hence called "nutricereals." Among many contributors to human nutrition, minerals play a significant part. They are indispensable for human growth and development and are divided into two groups—major and trace. The former constitutes sodium, potassium, calcium, phosphorus, magnesium, sulfur, and chloride which are required in large amounts and the latter consists of copper, fluoride, zinc, iron, chromium, selenium, iodine, molybdenum, and manganese, which are all required in small amounts. Interestingly, and despite the recent focus on different aspects of the nutritional value of minerals, evaluation of the mineral composition of millet grains has not yet been reported. Foxtail millet (cultivar 'RAU-8') has been reported to possess the maximum amount of seed proteins among all millets (Chandel et al. 2014). Furthermore, among millets, foxtail millet has one of the highest contents of protein, fat, ash, crude fiber, thiamine, and riboflavin (Table 10.1). The subsequent sections elaborate on the nutritional profile and associated benefits in foxtail millet as a "nutricereal." In millets, the grain is mostly eaten by humans and the leaves are mostly used for fodder. There are three major parts in grain, namely endosperm, germ, and bran; and each of these three parts is nutritionally different in the same crop as well as in different crops. Table 10.2 shows nutrient profiling of different parts of grain in two important millets—sorghum and pearl millet. Sorghum grain is rich in niacin, riboflavin, and pyridoxine, with negligible quantities of calcium and phosphorus. Pearl millet grain has very high calcium and phosphorus contents (Table 10.2).

10.2.1 Proteins

Proteins provide raw materials for general and essential amino acids. Millet grains are rich sources of seed storage proteins (SSPs) and are primarily used to facilitate seed germination following embryo development. Interestingly, they are rich in all essential amino acids except lysine and threonine, although sulfur-containing amino acids such as methionine and cysteine are present more abundantly. Their quantity and type are best suited for human consumption. The alcohol-soluble SSPs, namely the prolamins, constitute almost half the total protein content in millets wherein alpha-prolamins constitute a major component in all millets (Utsumi 1992). Within millets, proso millet and foxtail millet have the highest grain protein content and a recent study on the foxtail cultivar RAU-8 indicated that the crop has 13.1% more protein (Table 10.1) than other millets. Furthermore, foxtail millet has the highest concentration (milligrams per gram) of leucine and isoleucine and the second highest concentration of phenylalanine and valine among all the millets, and consistently higher amounts of the majority of essential amino acids when compared to rice and wheat (Table 10.3).

In several cereal grains, including millets, grain protein content is inversely proportional to grain yield, grain weight, and starch content and is positively correlated with ash content (Frey 1977; Subramanian and Jambunathan 1980). Grain protein is affected by the level of nitrogen fertilizer and higher nitrogen fertilizer increased grain protein content but render them with poor quality because of the increased accumulation of prolamin (Sawhney and Naik 1969).

Based on solubility, grain protein can be classified into four categories: (1) water soluble, albumin; (2) soluble in dilute salt solution, globulin, (3) alcohol soluble, prolamin, and (4) extractable in dilute alkali or acid solutions, glutenin. Different soluble fractions of important millets are given in Table 10.4. Pearl millet has a higher level of albumin; however, sorghum has a higher level of cross-linked prolamin, ß-prolamin.

10.2.2 Vitamins

Vitamins are an essential external dietary requirement and are critical for regulating human physiology. Millets are rich sources of the many vitamins, and the same is especially true for

Table 10.1 Approximate nutrient composition of millets and non-millet cereals (Saleh et al. 2013; Hulse et al. 1980; Ghosal and Krishna 1995)

Crop	Protein (g)	Fat (g)	Ash (g)	Crude fiber (g)	Carbohydrate (g)	Energy (kcal)	Ca (mg)	Fe (mg)	Thiamin (mg)	Riboflavin (mg)	Niacin (mg)
Rice (brown)	7.9	2.7	1.3	1	76	362	33	1.8	0.41	0.04	4.3
Wheat	11.6	2	1.6	2	71	348	30	3.5	0.41	0.1	5.1
Maize	9.2	4.6	1.2	2.8	73	358	26	2.7	0.38	0.2	3.6
Sorghum	10.4	3.1	1.6	2	70.7	329	25	5.4	0.38	0.15	4.3
Pearl millet	11.8	4.8	2.2	2.3	67	363	42	11	0.38	0.21	2.8
Finger millet	7.7	1.5	2.6	3.6	72.6	336	350	3.9	0.42	0.19	1.1
Foxtail millet	11.2	4	3.3	6.7	63.2	351	31	2.8	0.59	0.11	3.2
Common millet	12.5	3.5	3.1	5.2	63.8	364	8	2.9	0.41	0.28	4.5
Little millet	9.7	5.2	5.4	7.6	60.9	329	17	9.3	0.3	0.09	3.2
Barnyard millet	11	3.9	4.5	13.6	55	300	22	18.6	0.33	0.1	4.2
Kodo millet	9.8	3.6	3.3	5.2	66.6	353	35	1.7	0.15	0.09	2

Table 10.2 Nutrients composition of whole grain and its parts in millets

Grain/parts	Grain weight (%)	Protein (%)	As (%)	Oil (%)	Starch (%)	Ca (mg/kg)	P (mg/kg)	Niacin (mg/100 g)	Riboflavin (mg/100 g)	Pyridoxin (mg/100 g)
Sorghum (Hubbard et al. 1950)										
Whole grain	100	12.3	1.67	3.6	73.8	–	–	4.5	0.13	0.47
Endosperm	82.3	12.3	0.37	0.6	82.5	–	–	4.4	0.09	0.4
Germ	9.8	18.9	10.4	28.1	13.4	–	–	8.1	0.39	0.72
Bran	7.9	6.7	2	4.9	34.6	–	–	4.4	0.4	0.44
Pearl millet (Abdelrahman et al. 1984)										
Whole grain	100	13.3	1.7	6.3	–	55	358	–	–	–
Endosperm	75	10.9	0.32	0.53	–	17	240	–	–	–
Germ	17	24.5	7.2	32.2	–			–	–	–
Bran	8	17.1	3.2	5	–	168	442	–	–	–

Table 10.3 Essential amino acid profile of millets compared to fine cereals (mg/g of N)

Millets	Arginine	Histidine	Lysine	Tryptophan	P. Alanine	Tyrosine	Methionine	Cysteine	Threonine	Leucine	Isoleucine	Valine
Foxtail millet	220	130	140	60	420	–	180	100	190	1040	480	430
Proso millet	290	110	190	50	310	–	160	–	150	760	410	410
Finger millet	300	130	220	100	310	220	210	140	240	690	400	480
Little millet	250	120	110	60	330	–	180	90	190	760	370	350
Barnyard	270	120	150	50	430	–	180	110	200	650	360	410
Sorghum	240	160	150	70	300	180	100	90	210	880	270	340
Bajra	300	140	190	110	290	200	150	110	140	750	260	330
Rice	480	130	230	80	280	290	150	90	230	500	300	380
Wheat	290	130	170	70	280	180	90	140	180	410	220	280

Source FAO (1995)

Table 10.4 Protein fractions in millet grains (percent of total protein) (Jambunathan et al. 1984; Virupaksha et al. 1975; Monteiro et al. 1982)

Fraction	Sorghum		Pearl millet		Finger millet		Foxtail millet	
	Range	Mean	Range	Mean	Range	Mean	Range	Mean
Albumin + globulin	17.1–17.8	17.4	22.6–26.6	25	17.3–27.6	22.4	11.6–29.6	17.1
Prolamin	5.2–8.4	6.4	22.8–31.7	28.4	24.6–36.2	32.3	47.6–63.4	56.1
Cross-linked prolamin	18.2–19 5	18.8	1.8–3.4	2.7	2.5–3.3	2.78	6.4–17.6	8.9
Glutelin-like	3.4–4.4	4	4.7–7.2	5.5	–	–	5.2–11.9	9.2
Glutelin	33.7–38.3	35.7	16.4–19.2	18.4	12.4–28.2	21.2	–	6.7
Residue	10.4–10.7	10.6	3.3–5.1	3.9	16.1–25.3	21.3	–	2
Total	91.2–94.0	92.9	78.6–87.5	83.9	74.7–83.9	78.7	–	98

Table 10.5 Profile of vitamins (mg/100 g) among millets and non-millet cereals

Crop	Thiamine	Niacin	Riboflavin	Carotene	Vit B6	Folic acid	Vit B5	Vit E
Foxtail millet	0.59	3.2	0.11	32	–	15	0.82	31
Proso millet	0.41	4.5	0.28	0	–	–	1.2	–
Finger millet	0.42	1.1	0.19	42	–	18.3	–	22
Little millet	0.3	3.2	0.09	0	–	9	–	–
Barnyard millet	0.33	4.2	0.1	0	–	–	–	–
Kodo millet	0.15	2	0.09	0	–	23.1	–	–
Sorghum	0.38	4.3	0.15	47	0.21	20	1.25	12
Bajra	0.38	2.8	0.21	132	–	45.5	1.09	19
Rice	0.41	4.3	0.04	0	–	8	–	–
Wheat	0.41	5.1	0.1	64	0.57	36.6	–	–

foxtail millet grains. The crop has the highest concentrations of thiamine and vitamin E among millets and rice/wheat, with consistently higher amounts of other vitamins except for vitamin B6 (Table 10.5). However, millets lack vitamin A (although some foxtail millet and finger millet yellow endosperm contain small amounts of 13-carotene, a precursor of vitamin A) and vitamin C. Carotenoids isolated from sorghum are known to contain lutein, zeaxanthin, and ß-carotene ranging from 0 to 0.097 mg/100 g of grain (Blessin et al. 1958).

10.2.3 Macronutrients and Trace Elements

Phosphorus is an important structural component of all nucleic acids and cell membranes and is involved in many of the crucial signaling and regulatory processes along with calcium. Foxtail millet has the highest grain phosphorus content (422 mg/100 g) (Table 10.6). Millets are also good sources of trace elements. Recent studies on Zn and Fe content in millets have revealed that foxtail millet is one of the richer sources of these at 2.9 and 5.3 mg/100 g grain, respectively (National Institute of Nutrition 2007). It is not surprising that millets, in general, have consistently overall higher contents of major minerals and therefore stands to supplement our traditional food regimes efficiently.

10.2.4 Starch

Starch is the main carbon and energy source in the human diet (James et al. 2003). Two fractions of this component, namely amylose and amylopectin, get converted to simple sugar molecules following the digestive process. Based on the ease of digestion of starch in the human gut, starch is categorized as

Table 10.6 Micronutrient concentration in millets (mg/100 g dry mass)

Crop	P	Mg	Ca	Fe	Zn	Cu	Mn	Mo	Cr
Pearl millet	379	137	46	8.0	3.1	1.06	1.15	0.07	0.023
Finger millet	320	137	398	3.9	2.3	0.47	5.49	0.10	0.028
Foxtail millet	422	81	38	5.3	2.9	1.60	0.85	–	0.070
Proso millet	281	117	23	4.0	2.4	5.80	1.20	–	0.040
Little millet	251	133	12	13.9	3.5	1.60	1.03	–	0.240
Barnyard millet	340	82	21	9.2	2.6	1.30	1.33	–	0.140
Kodo millet	215	166	31	3.6	1.5	5.80	2.90	–	0.080

Table 10.7 Composition of soluble sugars in millets (in g/100 g dry weight)

Grain	Total sugar	Sucrose	Glucose + fructose	Raffinose	Stachyose
Sorghum	2.25 (1.3–5.2)	1.68 (0.9–3.9)	0.25 (0.06–0.74)	0.23 (0.10–0.39)	0.1 (0.04–0.21)
Pearl millet	2.56 (2.16–2.78)	1.64 (1.32–1.82)	0.11 (0.08–0.16)	0.71 (0.65–0.84)	0.09 (0.06–0.13)
Finger millet	0.65 (0.59–0.69)	0.22 (0.20–0.24)	0.16 (0.14–0.19)	0.07 (0.06–0.08)	–
Foxtail millet	0.46	0.15	0.1	0.04	–
Proso millet	–	0.66	–	0.08	–

rapidly digestible starch (RDS), slowly digestible starch (SDS), and resistant starch (RS) (Englyst et al. 1992). Furthermore, digestibility of starch depends on several factors such as the surface area of the starch particle and the ratio of amylase to amylopectin. Lower surface area and amylose content facilitate better digestibility of sorghum starch, and although the major non-millet seeds are rich in RDS, foxtail and other millets are surprisingly rich in RS and SDS. This has huge implications toward addressing and controlling the unusually higher GI contributed by RDS present in non-millet cereals. Supplementation of stable cereal with foxtail millets ensures slow, and consistent release of glucose, thereby preventing a postprandial spike in blood glucose, plasma cholesterol and triglyceride contents, increase in fat storage, and general insulin sensitivity (Higgins 2004). Examination of defatted white and yellow foxtail millet revealed their RS content to be as high as 13.35 and 14.56%, respectively (Bangoura et al. 2012; Fincher 1989). In sorghum, total starch content ranged from 56 to 73%; of this, 70–80% was amylopectin and 20–30% amylase (Jambunathan and Subramanian 1988; Deatherage et al. 1955). Waxy and sugary sorghum contain 0 and 5–15% amylose content, respectively (Ring et al. 1982; Deatherage et al. 1955; Singh and Axtell 1973). Water-soluble polysaccharides are also available in sorghum grains in sufficient quantity (21%). In pearl millet and finger millet, starch content ranged from 62.8 to 70.5% (Jambunathan and Subramanian 1988) and 59.5–61.25% (Wankhede et al. 1979), respectively. The composition of soluble sugar in millets is given in Table 10.7 (Subramanian et al. 1980, 1981; Murty et al. 1985; Becker and Lorenz 1978).

10.2.5 Lipids

Lipids serve important functions which involve major physiological, regulatory, and storage processes in the cell. Lipid fraction in seeds is primarily contributed by germ and aleurone layers (Rooney and Serna-Saldivar 1991). Foxtail millet is a rich source of stearic and linoleic acids, which are relatively sparse in other millet and non-millet cereals (Table 10.8). In an interesting study, fatty acid and lipid composition of glutinous and non-glutinous varieties of foxtail millet were compared to reveal that linoleic acid comprises almost 75% of the total fatty acids in them (Sridhar and Lakshminarayana 1992). It

Table 10.8 Major fatty acid composition of millets and non-millet cereals

Crop	Palmitic	Palmoleic	Stearic	Oleic	Linoleic	Linolenic
Foxtail millet	6.40	–	6.30	13.0	66.50	–
Proso millet	–	10.80	–	53.80	34.90	–
Finger millet	–	–	–	–	–	–
Little millet	–	–	–	–	–	–
Sorghum	14.0	–	2.10	31.0	49.0	2.70
Bajra	20.85	–	–	25.40	46.0	4.10
Rice	15.0	–	1.90	42.50	39.10	1.10
Wheat	24.50	0.80	1.00	11.50	56.30	3.70

was found that the main differences between glutinous and non-glutinous types are caused by differences in their stearic and arachidic acid compositions (Sridhar and Lakshminarayana 1992). Furthermore, foxtail millet bran oil is rich in palmitic (16.7%), oleic (24.7%), and linoleic (45.7%) acids (Krishnappa 2009).

10.3 Genetic Approaches for Micronutrient Enrichment in Crops

To provide for optimum dietary consumption of all important vitamins and minerals and also to increase the consumption of beneficial phytochemicals, scientists have turned to the study and manipulation of specific components of plant secondary metabolism. Such research is primarily aimed at augmenting the availability of the target compound within the staple crop up to the desired dietary levels by identification and suitable manipulation of relevant genes (Della Penna 1999). Such strategies rely heavily on the suitability of target compounds, their effectiveness, and whether their increased uptake will have any undesirable effects on humans. Interestingly, for specific vitamin (folate, vitamins E, B6, and A) and mineral (iron, calcium, selenium, and iodine) targets, the upper safe levels are much higher and 2–13 times that of RDA respectively, thereby allowing a greater window for manipulation (Lachance 1998; Della Penna 1999). Interestingly, provitamin A carotenoids (furnished by plants as β-carotene) has an upper safe level 100 times that of the RDA, thereby alleviating the risk of over-accumulation of vitamin A (retinol), for which the upper safe limit is only five times that of the RDA. Genetic manipulation strategies would, therefore, require augmenting provitamin A biosynthesis in plants rather than concentrating on vitamin A biosynthesis. Similar approaches may be undertaken in selecting other nutritionally important compounds for manipulation to minimize and entirely evade potentially adverse side effects of such technologies.

In addition to vitamins and minerals, many phytochemicals have health-promoting effects that are derived primarily from epidemiological studies, out of which a few have had their active chemical compound identified and are being rigorously tested. Glucosinolates are known to reduce carcinogen-DNA interaction and increase carcinogen detoxification (Hecht 1999) and isoflavones (phytoestrogens). Similar to genistein and daidzein, have been reported to reduce the incidence of many cancers, coronary heart diseases, and osteoporosis (Kurzer and Xu 1997). Given their evident health-promoting effects and technological feasibility toward manipulating their synthesis in staple food crops, efforts can now be initiated in this direction with significant advantages.

10.4 QTL/Genes for Nutrient Content in Millets and Cereals

Advancement in genotyping technologies such as next-generation sequencing and genotyping by sequencing and statistical tools facilitate the identification of QTLs/genes in several crops for each of the important traits (Huang et al. 2010; Jia et al. 2014; Jaiswal et al. 2016; Su et al. 2016). Using phenotyping and genotyping data, QTLs can be identified through different approaches such as QTL interval mapping and genome-wide association mapping. Genetic resources with high diversity for the trait of interest is essential to study the genetic architecture of the trait. QTLs for nutrient synthesis and accumulation in millet may prove to be vital assets for identification of candidate genes and further utilization in millet breeding for high nutrient value through marker-assisted selection, marker-assisted recurrent selection, or genomic selection. Unlike major cereals such as wheat and rice, millets are "orphan crops" with only a few studies undertaken to identify genes/QTLs nutrients in them (Table 10.9). In pearl millet, two colocalized QTLs for Zn and Fe were found on linkage group 3, using 106 recombinant

Table 10.9 QTL mapping for nutrient traits in important millets and cereals

Crop	Trait	Population	Approach	References
Rice	Zn	RILs, ILs	Interval mapping	Zhang et al. (2014)
	Zn	BILs	Interval mapping	Ishikawa et al. (2010)
	Zn	RILs	Interval mapping	Anuradha et al. (2012)
	Zn	DH	Interval mapping	Zhang (2011)
Wheat	Zn, Fe, Cu, Mn, Se	RILs	Interval mapping	Pu et al. (2013)
	Zn, Fe	RILs	Interval mapping	Crespo-Herrera et al. (2016)
	Zn, Fe	RILs	Interval mapping	Zhi-en et al. (2014)
	Zn, Fe	RILs	Interval mapping	Peleg et al. (2009)
	Zn, Fe, protein	RILs	Interval mapping	Xu et al. (2012)
Pearl millet	Zn, Fe	RILs	Interval mapping	Kumar et al. (2016)
Sorghum	Zn, Fe	Germplasm	GWAS	Kotla et al. (2015)
Foxtail millet	Agronomic traits	Varieties	GWAS	Jia et al. (2016)
	Branching	F_3	Interval mapping	Doust et al. (2004)
	Agronomic/yield traits	F_2	Interval mapping	Fang et al. (2016)
	Drought tolerance	RILs	Interval mapping	Qie et al. (2014)

inbred lines (RILs) derived from ICMB841-P3x8638-P2, with PVE (Phenotypic Variation Explained) 36 and 19%, respectively (Kumar et al. 2016). However, no major epistasis (QxQ, QxQxE) was observed for these QTLs. In the case of sorghum, Kotla et al. (2015) identified candidate genes for Zn and Fe which are homologous to nutrient (Zn, Fe)-related genes in rice and barley (OsNAS, OsZIP, OsNAC, HvNAAT, HvDMAS, etc.) through genome-wide association studies (GWAS). Unfortunately, in foxtail and kodo millets no studies to identify genomic regions for nutritionally important traits have been documented so far. In foxtail millet, comparative genomics may prove to be a useful approach to identify genomic regions associated with potential nutrients given the recent availability of its genome sequence (Muthamilarasan et al. 2016). Although QTL mapping/GWAS for micro/macro nutrient traits in the crop are currently unavailable, similar approaches have been undertaken for yield and agronomic traits (Doust et al. 2004; Qie et al. 2014; Fang et al. 2016). Similar approaches could be very useful in other millet crops with limited mapping populations and molecular marker densities.

10.5 Improving Plant Nutrient Contents Through Breeding Strategies

Notwithstanding the fact that molecular-genetic and other emerging technologies hold much promise in augmenting nutritional value of crops, conventional techniques are nevertheless important and should be implemented either in parallel or independent of emerging ones. Unfortunately, the focus of such strategies so far has been improving yield to meet the calorific needs of the increasing world population, and programmes aimed at improving the nutritional and micronutrient worth of the crops have been largely overlooked. Fortunately, substantial genetic variation exists between the genotypes with regard to occasional cases where micronutrients concentrations have been accessed (Schonhof and Krumbein 1996; Wang and Goldman 1996). These variations are now being used to obtain nutritionally enriched cultivars and can prove to be useful tools for identifying and establishing the much needed physiological and genetic basis for nutrient variation (Grusak et al. 1999).

10.6 Molecular Genetics and Genomics of Plant Secondary Metabolism and Crop Improvement

Previous efforts to modify plant secondary metabolisms suffered because of the dearth of mechanistic understanding of the intricate processes that regulate the same (Della Penna 1999). For many of the candidates that are of functional importance to humans, molecular genetic approaches have replaced the traditional ones in the identification of encoding and associated genes. For example, plants with altered production of tocopherols (Norris et al. 1995), carotenoids (Pogson et al. 1996), flavonoids (Shirley et al. 1995), and ascorbic acid (Conklin et al. 1997) have been obtained by such technologies to investigate their genetic basis. Heterologous bacterial systems have also been employed to characterize the role of plant enzymes in influencing the uptake of iron and synthesis of thiamine, biotin, and vitamin E (Norris et al. 1998; Baldet et al. 1997; Belanger et al. 1995; Eide et al. 1996).

A large number of ongoing sequencing projects in many different organisms present new opportunities for plant researchers to study plant metabolism in a more comprehensive manner. Out of many varieties of DNA sequences yielded in such projects, a small but substantial group encode novel (pioneer) sequences with no related orthologs in the database (Sterky et al. 1998; Yamamoto and Sasaki 1997). As a large number of secondary compounds are unique to plants, many pioneers and unknown plant sequences likely encode structural enzymes and regulatory components of plant secondary metabolism. Plant scientists have begun mining and understanding the huge genomic resources and microarray data which can help existing genetic and biochemical strategies for a better understanding of the complex metabolic pathways in plants.

10.7 Conclusions and Future Perspectives

Despite the obvious advantages nutritional genomics offers us, it is not without its share of limitations. For example, it is limited to the metabolic pathways (such as vitamins) for which information is available in model organisms but not applicable for other phytochemical pathways that are kingdom specific (Della Penna 1999). Moreover, extreme diversity of phytochemicals and their limited evolutionary distribution are major challenges toward gene identification via this approach. Furthermore, many plants that are being considered as genomic resources (e.g., Arabidopsis, maize, and rice) do not synthesize many of the useful health-promoting and well-characterized phytochemicals. It is against this backdrop that scientists restrict their research to a few species where the target compound flux is high. Moreover, if such plants are non-model, researchers have to confront the lack of genomic, genetic, and molecular tools otherwise available with model plants. Interestingly, such problems can be overcome by large-scale genome sequencing projects and microarray expression studies of non-model plants and their variants known to accumulate higher amounts of the target compound. This approach is advantageous because it allows us to use any type of plant/specific tissue from non-model plants to generate ESTs and subject them to a set of specific expression studies and bioinformatics analysis to identify the gene pool specific to the target pathway followed by its functional analysis using heterologous systems in a range of organisms. Moreover, the approach also allows for the identification of related non-pathway genes (involved in intermediary metabolism) that regulate accumulation or increased flux of the target compound (Della Penna 1999). In this manner, the profound diversity of metabolites, their flux, their biochemistry, and the cell biology of non-model plants can aid in obtaining a more comprehensive knowledge of a given pathway.

Acknowledgements Studies on millet genomics in Dr. Manoj Prasad's laboratory are supported by Science and Engineering Research Board (SERB), Department of Science and Technology (DST), Govt. of India [Grant No. EMR/2015/000464], by Department of Biotechnology, Govt. of India [Grant No. BT/HRD/NBA/37/01/2014], and by Core Grant of National Institute of Plant Genome Research (NIPGR), New Delhi, India.

References

Abdelrahman A, Hoseney RC, Varriano-Marston E (1984) The proportions and chemical compositions of hand-dissected anatomical parts of pearl millet. J Cereal Sci 2:127–133

Anuradha K et al (2012) Mapping QTLs and candidate genes for iron and zinc concentrations in unpolished rice of Madhukar×Swarna RILs. Gene 508:233–240

Austin DF (2006) Foxtail millets (*Setaria*: Poaceae) abandoned food in two hemispheres. Econ Bot 60:143–158

Baldet P, Alban C, Douce R (1997) Biotin synthesis in higher plants: purification and characterization of bioB gene product equivalent from *Arabidopsis thaliana* overexpressed in *Escherichia coli* and its subcellular localization in pea leaf cells. FEBS Lett 419:206–210

Bangoura ML et al (2012) Starch functional properties and resistant starch from foxtail millet [*Setaria italica* (L.) P. Beauv] species. Pak J Nut 11:821–830

Becker R, Lorenz K (1978) Saccharides in proso and foxtail millets. J Food Sci 43:1412–1414

Belanger FC, Leustek T, Chu B, Kriz AL (1995) Evidence for the thiamine biosynthetic pathway in higher-plant plastids and its developmental regulation. Plant Mol Biol 29:809–821

Bergamini N et al (2013) Minor millets in India: a neglected crop goes mainstream. In: Fanzo J, Hunter D, Borelli T, Mattei F (eds) Diversifying food and diets: using agricultural biodiversity to improve nutrition and health. Bioversity International, Rome, pp 313–325

Black RE et al (2013) Maternal and child nutrition study group: maternal and child undernutrition and overweight in low-income and middle-income countries. Lancet 382:427–451

Blessin CW, VanEtten CH, Wiebe R (1958) Carotenoid content of the grain from yellow endosperm-type sorghums. Cereal Chem 35:359–365

Chandel G, Meena RK, Dubey M, Kumar M (2014) Nutritional properties of minor millets: neglected cereals with potentials to combat malnutrition. Curr Sci 107:1109–1111

Conklin PL, Pallanca, Last RL, Smirnoff N (1997) L-Ascorbic acid metabolism in the ascorbate- deficient arabidopsis mutant *vtc1*. Plant Physiol 115:1277–1285

Crespo-Herrera LA, Velu G, Singh RP (2016) Quantitative trait loci mapping reveals pleiotropic effect for grain iron and zinc concentrations in wheat. Ann Appl Biol 169:27–35

Deatherage WL, McMasters MM, Rist CE (1955) A partial survey of amylose content in starch from domestic and foreign varieties of corn, wheat and sorghum and from some other starch-bearing plants. Trans Am Assoc Cereal Chem 13:31–42

Deaton A, Dreze J (2009) Food and nutrition in India: facts and interpretations. Available at: http://www.princeton.edu/deaton/downloads/FoodandNutritioninIndiaFactsandInterpretations.pdf

Della Penna D (1999) Nutritional genomics: manipulating plant micronutrients to improve human health. Science 285:375–379

Doust AN et al (2004) Genetic control of branching in foxtail millet. Proc Natl Aced Sci U S A 101:9045–9050

Dwivedi S et al (2011) Millets: genetic and genomic resources. In: Janick J (ed) Plant Breed Rev 35, Wiley, New Jersey, pp 247–375

Eide D, Broderius M, Fett J, Guerinot ML (1996) A novel iron-regulated metal transporter from plants identified by functional expression in yeast. Proc Natl Acad Sci U S A 93:5624–5628

Englyst HN, Kingman SM, Cummings JH (1992) Classification and measurement of nutritionally important starch fractions. Eur J Clin Nutr 46:S33–S50

Fang X et al (2016) A high density genetic map and QTL for agronomic and yield traits in Foxtail millet [*Setaria italica* (L.) P. Beauv.]. BMC Genom 17:336

FAO (1995) Sorghum and millets in human nutrition. Rome, Italy. http://www.fao.org/docrep/T0818e/T0818E00.htm

Fincher GB (1989) Molecular and cellular biology associated with endosperm mobilization in germinating cereal grains. Annu Rev Plant Physiol Plant Mol Biol 40:305–346

Frey KJ (1977) Proteins of oats. Z Pflanzenzucht 78:185–215

Ghosal A, Krishna O (1995) Millets of India. Navdanya Publishers, New Delhi

Grusak MA, Pearson JN, Marentes (1999) The physiology of micronutrient homeostasis in field crops. Field Crops Res 60:41–56

Hecht SS (1999) Chemoprevention in cancer by isothiocyanates, modifiers of carcinogen metabolism. J Nutr 129:768–774

Hegde PS, Rajasekaran NS, Chandra TS (2005) Effects of the antioxidant properties of millet species on oxidative stress and glycemic status in alloxan induced rats. Nutr Res 25:1109–1120

Higgins JA (2004) Resistant starch: metabolic effects and potential health benefits. J AOAC Int 87:761–768

Hirschi KD (2009) Nutrient biofortification of food crops. Ann Rev Nutr 29:401–421

Huang X et al (2010) Genome-wide association studies of 14 agronomic traits in rice landraces. Nat Genet 42:961–969

Hubbard JE, Hall HH, Earle FR (1950) Composition of the component parts of the sorghum kernel. Cereal Chem 27:414–421

Hulse JH, Laing EM, Pearson OE (1980) Sorghum and the millets: their composition and nutritive value. Academic Press, New York

Ishikawa S et al (2010) A quantitative trait locus for increasing cadmium-specific concentration in rice grain is located on short arm of chromosome 7. J Exp Bot 613:923–934

Jaiswal V et al (2016) Genome wide single locus single trait, multi-locus and multi-trait association mapping for some important agronomic traits in common wheat (*T. aestivum* L.). PLoS ONE 11:e0159343

Jambunathan R, Subramanian V (1988) Grain quality and utilization of sorghum and pearl millet. In: Proceedings of the international biotechnology workshop in biotechnology in tropical crop improvement, Patancheru, Inde, 12–15 Janvier 1987, p 133–139

Jambunathan R, Singh U, Subramanian V (1984) Grain quality of sorghum, pearl millet, pigeonpea and chickpea. In Achaya KT (ed) Proceedings of a workshop in interfaces between agriculture nutrition and food science, Patancheru, Inde, 10–12 Nov 1981, Université des Nations Unies, Tokyo, Japan, p 4760

James MG, Denyer K, Myers AM (2003) Starch synthesis in the cereal endosperm. Curr Opin Plant Biol 6:215–222

Jayachandran S, Pande R (2013) Why are Indian children shorter than African children? Available at: http://www.hks.harvard.edu/fs/rpande/papers/ Indianchildrenheight.pdf

Jia YH et al (2014) Molecular diversity and association analysis of drought and salt tolerance in *Gossypium hirsutum* L. germplasm. J Integr Agric 13:1845–1853

Kotla A, et al. (2015) Genome-wide association analysis for Fe and Zn concentration in sorghum grains identifies the potential candidate genes for sorghum biofortification. PAG XXIII, San Diego CA, 10–14 January 2015

Krishnappa M (2009) Breeding potential of selected crosses for genetic improvement of finger millet. SAT eJ 7:1–6

Kumar S et al (2016) Mapping quantitative trait loci controlling high iron and zinc content in self and open pollinated grains of pearl millet [*Pennisetum glaucum* (L.) R. Br.]. Front Plant Sci 7:1636

Kurzer MS, Xu X (1997) Dietary phytoestrogens. Ann Rev Nutr 17:353–381

Lachance PA (1998) Overview of key nutrients: micronutrient aspects. Nutr Rev 56:S34–S39

Lata C, Gupta S, Prasad M (2013) Foxtail millet: a model crop for genetic and genomic studies in bioenergy grasses. Crit Rev Biotechnol 33:328–343

Mal B, Padulosi S, Ravi SB (2010) Minor millets in South Asia: learning from IFAD-NUS Project in India and Nepal. Biodiversity Intl, Maccarese, Rome, Italy and M.S. Swaminathan Research Foundation, Chennai, India, pp 1–185

Monteiro PV, Virupaksha TK, Rajagopol Rao D (1982) Proteins of Italian millet: amino acid composition, solubility fractionation and electrophoresis of protein fractions. J Sci Food Agric 33:1072–1079

Murty DS et al (1985) Soluble sugars in five endosperm types of sorghum. Cereal Chem 62:150–152

Muthamilarasan M, Prasad M (2015) Advances in *Setaria* genomics for genetic improvement of cereals and bioenergy grasses. Theor Appl Genet 128:1–14

Muthamilarasan M, Dhaka A, Yadav R, Prasad M (2016) Exploration of millet models for developing nutrient rich graminaceous crops. Plant Sci 242:89–97

Norris SR, Barrette TR, DellaPenna D (1995) Genetic dissection of carotenoid synthesis in *Arabidopsis* defines plastoquinone as an essential component of phytoene desaturation. Plant Cell 7:2139–2148

Norris SR, Shen X, DellaPenna D (1998) Complementation of the *Arabidopsis pds1* mutation with the gene encoding p-hydroxyphenylpyruvate dioxygenase. Plant Physiol 117:1317–1323

Parameswaran K, Sadasivam S (1994) Changes in the carbohydrates and nitrogenous components during germination of proso millet (*Panicum miliaceum*). Plant Foods Hum Nutr 45:97–102

Peleg Z et al (2009) Quantitative trait loci conferring grain mineral nutrient concentrations in durum wheat×wild emmer wheat RIL population. Theor Appl Genet 119:353–369

Pogson B, McDonald KA, Truong M, Britton G, DellaPenna D (1996) Arabidopsis carotenoid mutants demonstrate that lutein is not essential for photosynthesis in higher plants. Plant Cell 8:1627–1639

Pu ZE et al (2013) Quantitative trait loci associated with micronutrient concentrations in two recombinant inbred wheat lines. J Integ Agri Adv 13:2322–2329

Qie L et al (2014) Mapping of quantitative trait locus (QTLs) that contribute to germination and early seedling drought tolerance in the interspecific cross *Setaria italic*×*Setaria viridis*. PLoS ONE 9:e101868

Ring SH, Akingbala JO, Rooney LW (1982) Variation in amylose content among sorghums. In: Rooney LW, Murty DS (eds) Proceedings of the international symposium on sorghum grain quality, Hyderabad, India, 28–31 Oct 1981, pp 269–279

Rooney LW, Serna-Saldivar S (1991) Sorghum. In: Lorenz KJ, Kulp K (eds) Handbook of cereal science and technology. Marcel Dekker, New York, pp 233–269

Saleh ASM, Zhang Q, Chen J, Shen Q (2013) Millet grains: nutritional quality, processing, and potential health benefits. Compr Rev Food Sci Food Saf 12:281–295

Sawhney SK, Naik MS (1969) Amino acid composition of protein fractions of pearl millet and the effect of nitrogen fertilization on its proteins. Indian J Genet Plant Breed 29:395–406

Schonhof I, Krumbein A (1996) Gehalt an wertgebenden Inhaltstoffen verschiedener Brokkolitypen (Brassicaoleracea var italica Plenck). Gartenbauwissenschaft 61:281–288

Shirley BW, Kubasek WL, Storz G, Bruggemann E, Koornneef M, Ausubel FM, Goodman HM (1995) Analysis of *Arabidopsi* mutants deficient in flavonoid biosynthesis. Plant J 8:659–671

Singh R, Axtell JD (1973) Survey of world sorghum collection for opaque and sugary lines. In: Inheritance and improvement of protein quality and content in sorghum, No. 10 Research Progress Report No. 10, pp l–18. Lafayette, Indiana, Etats-Unis, Department of Agronomy, Agricultural Experiment Station Purdue University; Washington DC, Etats-Unis, Agence pour le développement international

Singh P, Raghuvanshi RS (2012) Finger millet for food and nutrition security. Afr J Food Sci 6:77–84

Sridhar R, Lakshminarayana G (1992) Lipid class contents and fatty acid composition of small millets: little (*Panicum sumatrense*), kodo (*Paspalum scrobiculatum*), and barnyard (*Echinochloa colona*). J Agric Food Chem 40:2131–2134

Sterky F, Regan S, Karlsson J, Hertzberg M, Rohde A, Holmberg A, Amini B, Bhalerao R, Larsson M, Villarroel R et al (1998) Gene discovery in the wood-forming tissues of poplar: analysis of 5,692 expressed sequence tags. Proc Natl Acad Sci U S A 95:13330–13335

Su J et al (2016) Detection of favorable QTL alleles and candidate genes for lint percentage by GWAS in Chinese upland cotton. Front Plant Sci 7:1576

Subramanian V, Jambunathan R (1980) Traditional methods of processing of sorghum (*Sorghum bicolor*) and pearl millet (*Pennisetum americanum*) grains in India. Rep Intl Assoc Cer Chem 10:115–118

Subramanian V, Jambunathan R, Suryaprakash S (1980) Note on the soluble sugars of sorghum. Cereal Chem 57:440–441

Subramanian V, Jambunathan R, Suryaprakash S (1981) Sugars of pearl millet [*Pennisetum americanum* (L.) Leeke] grains. J Food Sci 46:1614–1615

Undernutrition contributes to half of all deaths in children under 5 and is widespread in Asia and Africa. UNICEF Data: monitoring the situation of children and women. Available at: http://data.unicef.org/nutrition/malnutrition. Accessed on 01 Apr 15

Utsumi S (1992) Plant food protein engineering. Adv Food Nutr Res 36:89–208

Virupaksha TK, Ramachandra G, Nagaraju D (1975) Seed proteins of finger millet and their amino acid composition. J Sci Food Agric 26:1237–1246

Wang M, Goldman IL (1996) Phenotypic variation in free folic acid content among F_1 hybrids and open-pollinated cultivars of red beet. J Am Soc Hort Sci 121:1040–1042

Wankhede DB, Shehnaj A, Raghavendra Rao MR (1979) Carbohydrate composition of finger millet (*Eleusine coracana*) and foxtail millet (*Setaria italica*). Qual Plant Foods Hum Nutr 28:293–303

White PJ, Broadley MR (2005) Biofortifying crops with essential mineral elements. Trends Plant Sci 10:586–593

Xu YF et al (2012) Molecular mapping of QTLs for grain zinc, iron and protein concentration of wheat across two environments. Field Crops Res 138:57–62

Yamamoto K, Sasaki T (1997) Large-scale EST sequencing in rice. Plant Mol Biol 35:135–144

Zhang X (2011) Identification of quantitative trait loci for Cd and Zn concentrations of brown rice grown in Cd-polluted soils. Euphytica 180:173–179

Zhang M et al (2014) Mapping and validation of quantitative trait loci associated with concentration of 16 elements in un milled rice grain. Theor Appl Genet 127:137–165

Zhao FJ, McGrath SP (2009) Biofortification and phytoremediation. Curr Opin Plant Biol 12:373–380

Zhi-en P et al (2014) Quantitative trait loci associated with micronutrient concentrations in two recombinant inbred wheat lines. J Integ Agri 13:2322–2329

Regulation of Development and Stress Response by miRNAs

11

Amita Yadav, Gunaseelen Hari-Gowthem, Mehanathan Muthamilarasan and Manoj Prasad

Abstract

MicroRNAs (miRNAs) are ~18 to 24-nucleotide non-coding RNAs, which regulate gene expression at the transcriptional and post-transcriptional levels through cleavage or translational inhibition of target mRNA. In addition to the key regulatory processes such as cellular, biological, and developmental functions, a considerable fraction of miRNAs have been shown to play important roles in abiotic and biotic stress responses, such as nutritional deficiency, drought, salinity, cold, heat, oxidative stress, and heavy metal. Being a naturally stress tolerant crop, foxtail millet has invited research on genome-wide identification and characterization of miRNAs for their role(s) in abiotic stress. In this context, the present chapter summarizes the classes of small RNAs with emphasis on miRNAs and provides a snapshot on their roles in stress regulatory machinery in foxtail millet.

11.1 Introduction

To circumvent environmental stresses, plants exhibit stress tolerance or stress avoidance through acclimation and adaptation mechanisms, which are evolved through natural selection (Yamaguchi-Shinozaki and Shinozaki 2006). Tolerance to abiotic stresses is a very complex phenomenon as it includes the intricate interactions between stress responsive elements and various molecular and biochemical factors affecting plant growth and development. Recently, the role of microRNAs (miRNAs) in regulating stress-responsive molecular machinery has been identified. miRNAs are ~18- to 24-nucleotide (nt) non-coding RNAs, which regulate the gene expression at transcriptional and post-transcriptional levels through cleavage or translational inhibition of target mRNA (Fig. 11.1). In plants, miRNA genes generate

Amita Yadav and Gunaseelen Hari-Gowthem are contributed equally.

A. Yadav · G. Hari-Gowthem · M. Muthamilarasan · M. Prasad (✉)
National Institute of Plant Genome Research (NIPGR), Aruna Asaf Ali Marg, New Delhi 110067, India
e-mail: manoj_prasad@nipgr.ac.in

M. Prasad (ed.), *The Foxtail Millet Genome*, Compendium of Plant Genomes,
DOI 10.1007/978-3-319-65617-5_11

primary miRNA transcripts (pri-miRNAs) through transcription, which are cleaved by Dicer-like1 (DCL1) to form stem-loop miRNA: miRNA* duplexes known as pre-miRNA (Bologna and Voinnet 2014). Subsequently, Argonaute (AGO) protein binds to mature miRNAs to assemble RNA-induced silencing complex (RISC) (Jones-Rhoades and Bartel 2004). Mature miRNAs and RISC bind to cognate target genes and either cleave the target mRNAs with near perfect complementarity or repress their translation with lower complementarity (Bartel 2004). In addition to the role of miRNAs in key regulatory processes such as cellular, biological, and developmental functions, a considerable number of miRNAs have been shown to play important roles in abiotic and biotic stress responses, such as nutritional deficiency (Fujii et al. 2005; Liang et al. 2010), drought (Zhou et al. 2010), salinity (Li et al. 2013), cold (Zhang et al. 2009), heat (Chen et al. 2012), oxidative stress (Sunkar et al. 2006), and heavy metal stress (Zeng et al. 2012). Increasing evidence has revealed that miRNA-mediated gene regulation plays a significant role in water stress regulatory networks (Khraiwesh et al. 2012). Advances in next generation sequencing (NGS) and high-throughput sequence analysis platforms have accelerated the studies on plant miRNAs, leading to their discovery and functional characterization in many plant species (Morozova and Marra 2008).

Till now, 337 mature miRNAs in *Arabidopsis thaliana*, 401 in *Populus trichocarpa*, 164 in *Nicotiana tabacum*, 713 in *Oryza sativa*, 16 in *Saccharum officinarum*, 241 in *Sorghum bicolor*, 42 in *Triticum aestivum*, and 321 in *Zea mays* have been reported (miRBase; Release 20, June 2013). However, very much less work has been done on foxtail millet in this regard. Foxtail millet (*Setaria italica* L.) is considered as a model crop because of its small genome size ($2n = 2x = 18$; ~515 Mb) with a small amount

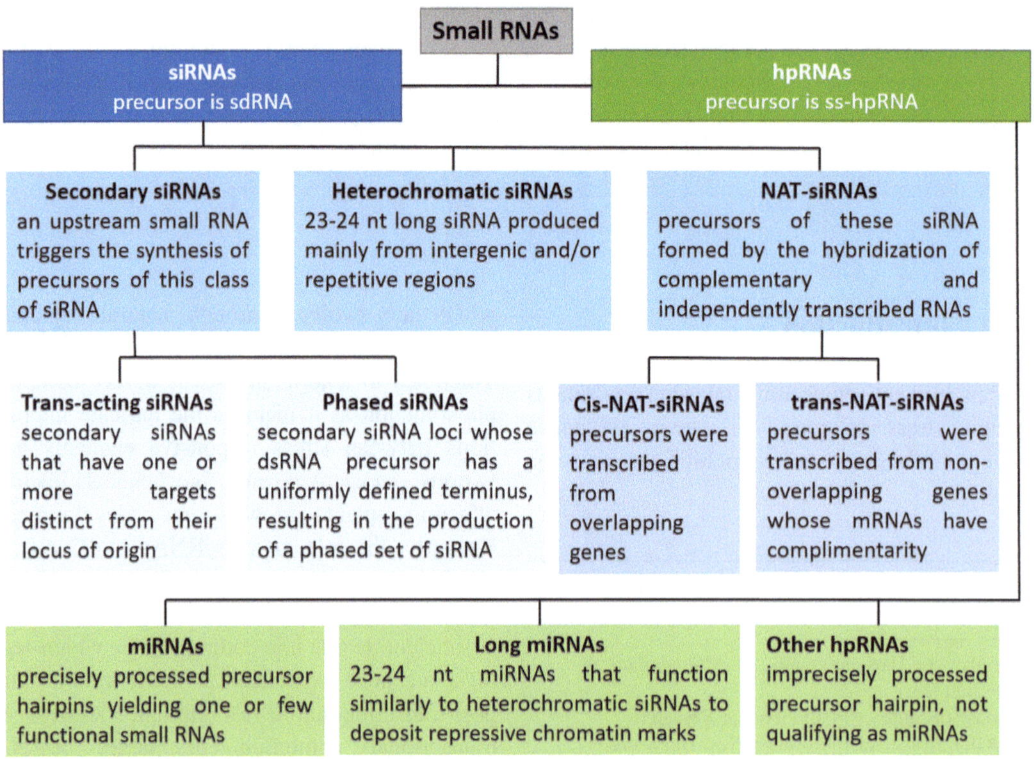

Fig. 11.1 Classification of endogenous plant small RNAs. dsRNA, double-stranded RNA; hpRNA, hairpin RNA; miRNA, microRNA; NAT-siRNA, natural antisense transcript small interfering RNA; siRNA, small interfering RNA

of repetitive DNA (30%), short growing duration, and self-pollinating and inbreeding nature (Muthamilarasan and Prasad 2015, 2017). It is an annual grass with thin, vertical, leafy stems, and erect and slender culms with hollow internodes. Moreover, its genome has now been sequenced by the Beijing Genomics Institute, China (Zhang et al. 2012) and the US Department of Energy-Joint Genome Institute (Bennetzen et al. 2012). The release of the genome sequences in public domain has resulted in the large-scale development of genetic and genomic resources. In view of this, the present chapter summarizes the role of miRNAs in regulating growth, development, and stress responsive molecular pathways, and enumerates the miRNAs identified and characterized in foxtail millet.

11.2 Discovery of miRNA

The first miRNA identified was lin-4 in *Caenorhabditis elegans* (Lee et al. 1993), which had a role in controlling the timing of larval development. Surprisingly, *lin-4* was not found to code for a protein but a pair of small RNAs—a shorter RNA which is about 22 nt in length, and a longer one approximately 61 nt long (stem-loop precursor). The lin-4 RNAs had antisense complementarity with multiple sites in 3′ UTR of *lin-14* gene, which significantly reduces the amount of LIN-14 protein without any considerable effect on *lin-14* mRNA level (Lee et al. 1993; Wightman et al. 1993). It was only after 2000 that this small regulatory RNA became complicated when *let-7* (22-nt regulatory RNA) was discovered in the same system, *C. elegans*, a gene that promotes the transition from late-larval to adult cell (Reinhart et al. 2000; Slack et al. 2000). Its homologs were also identified in the genomes of human, Drosophila, and bilateral animals (Pasquinelli et al. 2000). Over 100 additional genes were reported in various organisms such as flies, worms, and human cells in less than a year (Lagos-Quintana et al. 2001; Lau et al. 2001; Lee and Ambros 2001). Because of its developmental stage transition specific regulation, lin-4 and let-7 (founding members) were identified as small temporal RNAs (stRNAs). It is after this that the term ‘microRNA’ attracted attention, which included all other small RNAs with similar features (Lagos-Quintana et al. 2001; Lau et al. 2001; Lee and Ambros 2001).

11.3 Mechanism of miRNA Biogenesis and Their Mode of Action

Transcription, processing, modification, and finally loading to RISC are the steps in miRNA biogenesis. DNA-dependent RNA pol II transcribes a miRNA gene (MIR) into primary-miRNA (pri-miRNA) with a 5′ capping and 3′ polyadenylation. DCL 1 (RNAse III) helps in generating the hairpin structured precursor-miRNA (pre-miRNA) (Park et al. 2002; Reinhart et al. 2002) and the double-stranded miRNA-miRNA* duplex is formed by the endonucleolytic activity of DCL1, and DCL requires the proteins hyponastic leaves 1 (HYL1-double-stranded binding protein) and serrate (SE-Zn finger protein) for processing (Han et al. 2004; Vazquez et al. 2004; Lobbes et al. 2006; Yang et al. 2006) and Dawdle (DDL-nuclear RNA binding protein) promotes this step (Yu et al. 2005) and the stability of the duplex is brought about by methylation at 2′ OH of 3′ terminal nucleotides by Hua Enhancer 1 HEN1 (Yu et al. 2005). Following this, HASTY5 (HST5) export factor exports the modified miRNA-miRNA* duplex from nucleus to cytoplasm (Park et al. 2002) and the duplex is unwound in cytoplasm and one strand of duplex is loaded into AGO1 protein to form miRISC (Baumberger and Baulcombe 2005; Qi et al. 2005). Mature miRNA in RISC complex regulates the gene expression either by cleaving the transcript or inhibiting the translation by binding to the target sequence with near perfect complementarity (Chen 2004; Gandikota et al. 2007). Translation inhibition of gene occurs in the endoplasmic reticulum. Altered Meristem Program1 (AMP1) has been reported for interacting with AGO1 to facilitate the translational inhibition (Li et al. 2013).

11.4 Regulation of miRNA Accumulation and Activity

miRNAs are important regulators of gene expression at transcriptional as well as post-transcriptional level, with accumulation and activity of miRNA being the two deciding factors tuned when required, such as at different developmental stages and with environmental stress response.

11.4.1 Regulation of miRNA Biogenesis

11.4.1.1 Transcriptional Regulation of MIR Genes

MIR genes are usually located in intergenic regions. However, a few reports show they are present in the intronic or exonic region of host genes as well as in transposable elements (Xie et al. 2005; Piriyapongsa and Jordan 2008; Nozawa et al. 2012; Yang et al. 2012). At-negative on TATA less2 (NOT2) and cell division cycle 5 (CDC5) are a few transcription factors (TFs) required for efficient transcription of MIR genes. In addition to the above two TFs, specific TFs such as Powerdress (PWR) for *MIR172* family which helps in recruitment of Pol II to promoters (Yumul et al. 2013), TF Fused in Sarcoma 3 (FUSCA3) for *MIR156A* and *MIR156C* regulation (Wang and Perry 2013), and Apetala2 (AP2) TF for regulation of MIR156 and MIR172 genes in the opposite manner by promoting the expression of *MIR156* and repressing the transcription of *MIR172* (Yant et al. 2010) have been reported. In addition to TFs, molecules such as mediator complexes are essential in the regulation of MIR genes. Lack of mediator fades out the promoter activity of the MIR genes, resulting in suppressed levels of pri-miRNAs and miRNAs (Kim et al. 2011). Mediators form an essential link between a TF and RNA Pol II and are considered as TF regulators.

11.4.1.2 Effect of Splicing on Pri-miRNA Processing

Apart from the intergenic location of plant miRNAs, a significant number of MIR genes reside in the intronic region of protein-coding genes. In such a situation, transcriptional regulation of these host genes has an effect on the generation of pri-miRNAs because of splicing of the intron from which miRNA originates, which in turn affects the formation of stem-loop structure blocking the biogenesis of mature miRNAs. For example, heat stress processing of pri-miR400 is hindered because of alternate splicing of *At1g32583* which hosts the gene *MIR400;* however, in the absence of heat stress, pri-miR400 is normally processed to generate mature miRNA (Yan et al. 2012). Other such examples are *MIR162a*, *MIR842*, and *MIR846* in *Arabidopsis* (Hirsch et al. 2006; Jia and Rock 2013).

11.4.1.3 Regulation of DCL1 Activity

Following transcription is the processing of pri-miRNA into pre-miRNA by DCL1 in the nucleus, which depends upon the expression level, localization, and phosphorylation and dephosphorylation state of DCL1. Its activity is greatly influenced by the secondary structure of pri-miRNAs. In plants, the hairpin structure of pri-miRNAs is very heterogeneous in nature in respect of length, structure, and positioning of the miRNA/miRNA* duplex. Most of the multi-branched terminal loops containing pri-miRNAs prefer the base to loop processing for the efficient production of mature miRNA, which is catalyzed by DCL1. On the other hand, loop to the base processing of pri-miRNA transcripts having multi-branched terminal loops suppresses the generation of functional mature miRNA (Zhu et al. 2013). Exceptionally, pri-miR159a and pri-miR319a prefer the loop to base processing, probably because of their unusually long upper stem region (Bologna et al. 2009; Cuperus et al. 2010).

DCL1 is assisted by any proteins such as HYL1 (Yang et al. 2010), SE (Machida et al. 2011), tough (TGH) (Ren et al. 2012), CDC5 (Zhang et al. 2013b), stabilized1 (STA1) (Ben et al. 2013), DDL and C-terminal domain phosphatase-like 1 (CPL1) (Manavella et al. 2012a) for its precise functioning whereas it is negatively regulated by miR162 (Xie et al.

2003), a histone transferase (GCN5) (Kim et al. 2009), and short interspersed elements (SINEs). HY1 is a dsRNA-binding protein and probably binds to a double-stranded miRNA/miRNA* region as a dimer to facilitate the proper processing of pri-miRNAs; SE is a single-stranded RNA binding domain which helps in positioning of miRNA precursor toward the catalytic site of DCL1; TGH is a G-patch domain protein probably having a role in recruiting pri-miRNAs to DCL1, thus promoting the cleavage efficiency; interaction with CDC5 improves the activity of DCL1; STA1 is a pre-mRNA processing factor which regulates the transcription of *DCL1* in a positive manner; DDL interacts with negatively charged (phosphorylated) phosphothreonine motif present in the DCL1; CPL1 maintains the hypophosphorylated state of HYL1 by dephosphorylation for its optimal activity; DCL1 is negatively regulated by miR162 which balances the level of *DCL1* post-transcriptionally, thus affecting the processing of all pri-miRNAs; GCN5 negatively regulates the transcription of *HYL1* and *SE*; and SINEs sequester the *HYL1* from processing of pri-miRNA to pre-miRNA by mimicking the pri-miRNA structure.

11.5 miRNA Stability and Degradation

11.5.1 Methylation and Uridylation of miRNAs

HEN1, a methyltransferase enzyme, protects the miRNA/miRNA* duplex (Li et al. 2005; Yu et al. 2005) whereas its degradation is brought about by HESO1, a terminal uridyl transferase HEN1 suppressor1 (Ren et al. 2012); thus, the optimum level of miRNA is achieved by maintaining the balance between miRNA stability and degradation. Methylation is essential in preventing the miRNAs from uridylation activity of HESO1 (Ren et al. 2014).

11.5.2 Degradation of miRNAs by Exoribonucleases in Plants

In *Arabidopsis*, small RNA degrading nucleases (SDN1, SDN2, SDN3), which act as 3′-5′-exoribonucleases, act only upon 2′-O-methylated miRNAs, unlike HESO1 which acts on 3′-uridylated miRNAs (Ramachandran and Chen 2008); thus SDN1 and HESO1 probably cooperate in degrading miRNAs.

11.6 Regulation of miRNA Activity

11.6.1 Loading of AGO Proteins

AGO1 is a major effector protein in plants which consists of four functional domains named the N-terminal domain, the PAZ domain, the middle (MID) domain, and the PIWI domain. The loading of AGO1 is determined by several protein factors, miRNA/miRNA* duplex and 5′ nucleotides (Mi et al. 2008; Montgomery et al. 2008); squint (SQN), cyclophilin 40 (CyP40), and heat shock protein 90 (HSP90) are the protein factors which assist in loading (Smith et al. 2009; Iki et al. 2010, 2012; Earley and Poethig 2011). miRNA strand, having lower 5′ end thermostability, is the preferred strand for loading into AGO1 complex (Manavella et al. 2012b).

11.6.2 Competition of miRNAs with Endogenous RNAs

Cleavage of the transcript in the miRNA-guided regulation of gene expression is based on the near perfect complementation of miRNA and mRNA (Bartel 2004). In *Arabidopsis*, a non-coding RNA called Induced by Phosphate Starvation 1 (*IPS1*) contains the motif sequence complementary to miR399. This complementation is resistant to cleavage because of the formation of a mismatch loop at the cleavage site, resulting in the sequestration of the miR399, thus causing accumulation of the miR399 target *PHO2* in *IPS1* overexpressing lines (Franco-Zorrilla et al. 2007).

11.6.3 Feedback Regulation of MIR Genes Through Methylation

miRNA regulation is in turn regulated by DNA methylation of the MIR genes as a feedback. In rice, a 24-nt miRNA processed by DCL3 and associated with AGO4 directs the DNA methylation at MIR genes (Wu et al. 2010), and a 24-nt siRNA causes RNA-directed DNA methylation at cytosine with the help of AGO4 (Bologna and Voinnet 2014). The mechanism of miRNA-directed DNA methylation may possibly be similar to that of siRNAs.

11.7 Roles of miRNAs

11.7.1 Developmental Processes

miRNAs are conserved throughout the plant kingdom and play an important role in regulating various developmental pathways, evident from the pleiotropic developmental defects which were seen in *Arabidopsis* mutants (*dcl1*, *hyl1*, *se*, *hen1*, *ddl*, *ago1*) of miRNA biogenesis genes (Ramachandran and Chen 2008).

In the case of phase transition, overexpression of a well-conserved miR156 causes prolonged expression of juvenile characteristics and thus delayed flowering in *Arabidopsis* because miR156 targets 11 SPL genes, among which *SPL3*, *SPL4*, and *SPL5* are responsible for the vegetative phase to floral phase transition, while *SPL9 and SPL15* govern the plastochron length in *Arabidopsis* (Schwab et al. 2005; Wu and Poethig 2006; Wang et al. 2008). Another well-documented miRNA is mi172, which controls flowering time by regulating AP2-like genes, *Target Of Eat1* (*TOE1*), and *Target Of Eat2* (*TOE2*) in *Arabidopsis* (Aukerman and Sakai 2003) and *Glossy15* (*gl15*) in maize (Lauter et al. 2005). *Arabidopsis* plants overexpressing miR172 have shown early flowering and disrupted floral organ identity by repressing *TOE1* and *TOE2* (Aukerman and Sakai 2003). Similarly, in maize the vegetative phase change is regulated by the opposing actions of *gl15* and *miR172* (Lauter et al. 2005). Chuck et al. (2007) reported long vegetative phase and delayed flowering in *Corngrass1* (*Cg1*) mutant where miR156 was up-regulated, and miR172 was down-regulated, suggesting the coordinated behaviour of miR156 and miR172 in the regulation of vegetative and floral transition.

As far as hormone biosynthesis and signalling are considered, gibberellic acid controls developmental processes such as male fertility and flowering during short days which is brought about by the GAMYB gene family which is the target of miR159 (Achard et al. 2004; Millar and Gubler 2005). miR159 overexpression lines showed hyposensitivity to abscisic acid (ABA) (Reyes and Chua 2007). Auxin response factors (ARFs) which are involved in the auxin signaling pathway are the targets for miRN160 and miR167. The miR160-resistant *ARF10*, *ARF16*, and *ARF17* plants showed the pleiotropic developmental defects in the aerial part and root (Mallory et al. 2005; Wang et al. 2005; Liu et al. 2007). miR164 plays a role in mediating the auxin signaling cascade during the emergence of lateral root by targeting *NAC1* (Guo et al. 2005). *TIR1*, an auxin receptor, and related F-box genes are targeted by miRNA, miR393 affecting the auxin signaling pathway (Jones-Rhoades and Bartel 2004). Jasmonic acid (JA) biosynthesis is regulated by cleaving of Teosinte branched1/ Cycloidea/Proliferating cell factor1 (*TCP4*) by miR319; TF TCP4 binds to the promoter sequence of *Lipoxigenase2* (*LOX2*), a chloroplast enzyme, which catalyzes the initial step in the JA biosynthesis pathway (Schommer et al. 2008).

In the case of pattern formation and morphogenesis, miR164 targets *CUC1* and *CUC2* and regulates the establishment of organ boundaries, floral patterning, and leaf morphogenesis (Laufs et al. 2004; Mallory et al. 2004a; Baker et al. 2005; Nikovics et al. 2006; Sieber et al. 2007). miR165/166 regulates the development of lateral organs by targeting homeodomain-leucine zipper TFs in maize and *Arabidopsis.* Higher accumulation of miR165/166 targets on the adaxial surface to promote lateral organ identity, which is brought about by the enrichment of these miRNAs in the abaxial side of the leaf primordia (McConnell et al. 2001; Emery et al. 2003; Juarez et al. 2004; Mallory et al. 2004b). In *Arabidopsis*, miR172 represses expression of *AP2*, which is responsible for the identities of the various floral whorls (Aukerman and Sakai 2003; Chen 2004). The pattern formation during stomatal development in the *Brassicaceae* family was found to be regulated by miR824, which targets *AGL16*, an MADS-box gene (Kutter et al. 2007).

In the *leguminaceae* family the symbiotic association between the leguminous plant and rhizobial bacteria results in biological nitrogen fixation which is caused by the exchange of chemical factors (Cooper 2007). Inoculation with *Bradyrhizobium japonicum*, a nitrogen-fixing bacterium in the soybean roots, caused a differential expression of various miRNAs such as miR159 and miR393, which were up-regulated, and miR160 and miR1693 3 h post infection (phi) and miR168 and miR172 which were down-regulated for 12 phi after 3 h of up-regulation (Subramanian et al. 2008). Following rhizobial infection, the expression levels of miR160, miR393, miR164, and miR168 have changed, suggesting a connection between nodulation signaling and auxin homoeostasis through their targets such as *ARFs* (miR160), *TIR1* (miR393), *NAC1* (miR164), and *AGO1* (miR168) (Li et al. 2010a, b).

11.7.2 Biotic Stresses

Biotic factors such as bacteria, viruses, fungi, insects, and nematodes cause a heavy loss to the crop yield. Many miRNAs are involved in biotic stress-associated defense mechanisms to combat the pathogen infection to withstand the epidemic. In the case of bacterial infections, miR393 was found to provide basal resistance against Flg22, a bacterial-derived pathogen-associated molecular pattern (Navarro et al. 2006). Along with miR393, miR160, miR167 and miR159 targeted ARF family members to initiate defense mechanism against *Pseudomonas syringae* infection (Rhoades et al. 2002; Zhang et al. 2011). *Xanthomonas axonopodis* pv. *Manihotis* infection in Cassava induced the expression of miR160, miR167, miR393, and miR390 and suppressed the abundance of miR535, miR395, miR482, miR397, miR398, and miR408 (Perez-Quintero et al. 2012).

In the case of fungal infections, RNA silencing *Arabidopsis* mutants such as *sgs2*, *sgs3*, *ago7*, *dcl4*, *nrpd1a*, and *rdr2* were reported to be susceptible to Verticillium strains (Ellendorff et al. 2009). Up-regulation of miR393, miR444, and miR827 and down-regulated expression patterns of miR156, miR159, miR164, and miR396 were reported in common powdery mildew (*Erysiphe graminis* f.sp. *tritici*)-infected wheat leaves (Xin et al. 2010). Two novel miRNAs, m0001 and m0002, are reported to be involved in resistance to *Verticillium dahliae* infection in eggplant (Yang et al. 2013). Microarray analysis post-infection with fungus *Exserohilum turcicum* in maize showed a higher abundance of miR811 and miR845, suggesting its involvement in better resistance to *E. turcicum* infection (Wu et al. 2014).

Viral infections are tackled by the host miRNAs in plants. miR156, miR160, and miR164 get up-regulated upon viral infection in tobacco (Navarro et al. 2008), and significant higher expression of two miRNAs, bra-miR158, and bra-miR1885 was observed in *Brassica rapa* after infection with *Turnip mosaic virus*. Surprisingly, the predicted target of bra-miR1885 was the TIR-NBS-LRR class of disease-resistant proteins which requires additional investigation to understand the molecular mechanism related to viral infection.

11.7.3 Abiotic Stresses

Understanding the molecular basis of complicated genetic interactions in response to abiotic stress factors (drought, salinity, heat, cold, etc.) is necessary to develop what is necessary to face the challenges posed by climate change. Not only TFs, signalling components, and alterations in the expression level of a gene but also a significant number of miRNAs have been reported during various abiotic stress conditions (Zhang 2015). Changes in relative expression of miR396, miR168, miR167, miR165, miR319, miR159, miR394, miR156, miR393, miR171, miR158, and miR169 were observed in *Arabidopsis* facing drought stress (Liu et al. 2008). miR169 was down-regulated, which increased the abundance of its target nuclear factor YA5, which in turn induced various drought-responsive genes responsible for drought tolerance. Genome-wide expression analysis in rice revealed that a set of 16 miRNAs were down-regulated and 14 miRNAs were significantly up-regulated during drought stress (Zhou et al. 2010). Another study showed strong inverse correlation for miR398a/b and miR408 and their respective targets, namely copper superoxide dismutase (CSD1/2), mitochondrial cytochrome c oxidase, and plastocyanin, which suggests an important link in adaptation to drought and copper homoeostasis (Trindade et al. 2010). During drought stress, 13 miRNAs were observed to be differentially expressed in wild emmer wheat, which is claimed to be drought tolerant (Kantar et al. 2010). In soybean, miR166-5p, miR169f-3p, miR1513c, miR397ab, miR-Seq13, and miR166f showed differential expression in sensitive and tolerant cultivar, of immense importance when it comes to crop improvement (Kulcheski et al. 2011).

In *Arabidopsis*, salinity stress up-regulated miRNAs such as miR156, miR158, miR159, miR165, miR167, miR168, miR169, miR171, miR319, miR393, miR394, miR396, and miR397, whereas miR398 was found to be majorly down-regulated (Liu et al. 2008). A study involving contrasting maize lines, that is, salt tolerance (NC286) and salt sensitive (Huangzao4), revealed the differential expression of various miRNAs during salinity stress in both the cultivars, which could be used in priming salinity sensitive elite lines (Ding et al. 2009).

Temperature fluctuation in the surrounding environment of a plant requires a reprogramming of gene expression profile to maintain homoeostasis at the cellular, molecular, and physiological levels. Average growing season temperature has been elevated in the past few years, which adversely affects the plant's physiology such as seed maturation and grain filling. Nine heat stress-responsive miRNAs were identified in wheat, among which eight conserved miRNAs (miR156, miR159, miR160, miR166, miR168, miR169, miR393, and miR827) were up-regulated and only miR172 showed down-regulation after heat stress (Xin et al. 2010). Comparative investigation in heat-tolerant and heat-susceptible cultivars of wheat revealed the differential expression pattern of many miRNAs in both the cultivars at 40 °C heat stress (Xin et al. 2011). miRNAs have also been reported to play a major part in the regulatory network of cold-responsive genes, namely miR165/166, miR393, miR396, and miR408 in *Arabidopsis* (Sunkar and Zhu 2004; Liu et al. 2008). Conflicting expression patterns of some of the miRNAs, such as up-regulation of miR168 expression in poplar and *Arabidopsis* (Lu et al. 2008; Liu et al. 2008) compared to down-regulation of miR168 in rice (Lv et al. 2010) were also reported, and an opposite expression of miR171 was observed in *Arabidopsis* and rice (Liu et al. 2008; Lv et al. 2010).

11.7.4 Nutrient Homeostasis

Phosphate (Pi) is an essential nutrient which is involved in the vital biochemical reactions in the plants, such as DNA replication, phospholipid bilayer formation, and ATP synthesis. Pi starvation up-regulated the expression of miR156, miR399, miR778, miR827, and miR2111 whereas miR169, miR395, and miR398 were found to be down-regulated (Hsieh et al. 2009). Upon Pi starvation, phosphate starvation response 1 (PHR1) is induced, which is a TF and is the master regulator of the PHR regulation pathway in *Arabidopsis*; it induces the expression of miR399 which in turns down-regulates miR399 target PHO2 (E2 ubiquitin-conjugating enzyme), particularly in roots (Aung et al. 2006; Bari et al. 2006). Additionally, miR827 and miR2111 were up-regulated under Pi starvation, and target E3 ligases (At1g02860 and At1g63010, respectively), which indicates phosphate homeostasis is regulated by an miRNA-regulated ubiquitination-mediated pathway (Fujii et al. 2005; Chiou et al. 2006).

Another important macronutrient for plants is sulfur, an integral part of the amino acid residue cysteine, which is further converted into glutathione, phytoalexins, and glucosinolates; these are considered as sulfur-containing defense compounds. MiR395 induces gene families, sulfate transporter (SULTR2;1/AST68), and the members of the ATP sulfurylase family (APS1, APS3, and APS4) at sulfate deprivation (Jones-Rhoades and Bartel 2004; Allen et al. 2005). Remobilizing the sulfate between leaves during sulfate scarcity is believed to be brought about by miR395 in *Arabidopsis* (Liang et al. 2010). These miRNAs help the plant in acclimatizing growth and development in a situation of sulfate scarcity.

11.7.5 Oxidative Stress and Hypoxia

Oxidative damage of the cell is caused by the toxic compounds called reactive oxygen species (ROS) produced as by-products of photosynthesis and respiration (Mittler et al. 2004); however, ROS also play a vital role in the regulation of various biological processes (Mittler et al. 2004). ROS is also a secondary messenger molecule in stress-associated signal transduction pathways (Vandenabeele et al. 2003; Mittler et al. 2004). In *Arabidopsis*, oxidative stress suppresses the expression of miR398 which otherwise down-regulates *CSD1* and *CSD2* (Cu–Zn superoxide dismutases) (Bonnet et al. 2004; Jones-Rhoades and Bartel 2004). miR398-resistant transgenic lines are more tolerant to high-intensity light, heavy metals, and other oxidative stressors because of the accumulation of *CSD2* (Sunkar et al. 2006).

Hypoxia is the situation of low oxygen leading to switching from aerobic to anaerobic respiration (Agarwal and Grover 2006; Bailey-Serres and Voesenek 2008). Zm-miR166, Zm-miR167, Zm-miR171, Os-miR396, Zm-miR399, Zm-miR159, At-miR395, Pt-miR474, and Os-miR528 were identified using microarrays in submerged maize roots (Zhang et al. 2008). Nineteen hypoxia-responsive miRNA families in *Arabidopsis* roots were identified using next-generation sequencing, which acted inversely with their corresponding targets (Moldovan et al. 2009).

11.7.6 Mechanical Stress

Bending of branches and stems by gravity, wind, or other external forces causes mechanical stress to plants. Transcript levels of miRNAs in tension-stressed compression-stressed xylem and unstressed xylem were compared to find reduced expression of some miRNAs (miR156, miR162, miR164, miR475, miR480, and miR481) and induced expression of miR408 at stress. However, miR160 and miR172 were down-regulated specifically in compression stress, and miR168 was up-regulated in the tension-created tissue (Lu et al. 2005).

11.7.7 ABA-Mediated Stress Responses

ABA is an important phytohormone which mediates the expression of stress-associated genes at dehydration (Koornneef et al. 1998; Wilkinson and Davies 2002). Increased sensitivity of the mutants of HYL1, DCL1, HEN1, SE, and HASTY genes sheds light on the involvement of miRNAs in ABA-mediated response in *Arabidopsis* (Lu and Fedoroff 2000; Zhang et al. 2008). ABA treatment induced the expression of miR159 during seed germination (Reyes and Chua 2007). It also induced miR393, miR397b, and miR402 and suppressed miR398a (Sunkar and Zhu 2004) in *Arabidopsis*, and up-regulated miR159.2, miR393, miR2118, miRS1, miR1514, and miR2119 in *Phaseolus vulgaris* (Arenas-Huertero et al. 2009).

11.8 Impact of miRNAs on Abiotic Stresses

A wide regulatory function of the miRNAs is in different developmental stages and at different stress, and is caused by evolutionary conservation and abundance. Various studies have over-expressed abiotic stress responsive miRNA genes such as miR156, miR159, miR169, miR319, miR393, miR394, miR395, miR396, miR402, miR417, and miR828 in different plant species. These miRNA-overexpressing transgenic plants were found to be tolerant to harsh environmental conditions. miR156 was found to regulate the recurring heat stress memory in *Arabidopsis*, and these plants were found to be more tolerant to heat stress (Stief et al. 2014); the same miRNA over-expressed in switchgrass produced higher biomass as compared to the control plant (Fu et al. 2012). In contrast, miR159 over-expressing rice plants were more sensitive to heat stress (Wang et al. 2012). Another conserved miRNA, miR169, is considered as a promising candidate for enhancing tolerance to drought and nitrogen deficiency conditions caused by reduced stomatal aperture index, stomatal conductance, and transpiration rate as compared to non-transgenic tomato plants, thus reducing water loss through transpiration and enabling survival when water is scarce (Zhang et al. 2011). Simultaneously, over-expression of miR169 caused hypersensitivity to nitrogen starvation, leading to yellowing of leaves (Zhao et al. 2011), and constitutive expression of miR169 in potato increased water retention and cell membrane integrity (Pieczynski et al. 2013). Expression of miR319 was found to regulate metabolism at multiple stresses just as over-expression of miR319 in creeping bentgrass conferred tolerance to salinity and drought stress (Zhou et al. 2013) and rice transgenics overexpressing miR319 were found to be tolerant to cold stress (Yang et al. 2013).

Transgenics with constitutive expression of miR395 (Kim et al. 2010a) and miR393 (Gao et al. 2011) in rice and *Arabidopsis* reduced the seedling growth and root development, respectively, and increased sensitivity to drought and salt stress in both the cases; but transgenic rapeseed plants overexpressing miR395 expressed significant enhancement of tolerance to cadmium stress (Zhang et al. 2013a). Over-expression of miR396 showed stunted root growth and decreased plant growth and development, as it negatively regulated the responses to salinity and alkalinity (Gao et al. 2010). In *Arabidopsis*, salinity, dehydration, and cold stresses up-regulated miR402 and enhanced plant growth has been observed under salinity stress only in transgenic plants overexpressing miR402 (Sunkar and Zhu 2004; Kim et al. 2010b). Wound-induced miRNA, miR828, showed higher lignin biosynthesis and H_2O_2 production in miR828 overexpressing lines in sweet potato, playing a role in defense mechanisms (Lin et al. 2012).

11.9 miRNAs of Foxtail Millet and Their Roles in Diverse Molecular Mechanisms

Compared to other crops, foxtail millet has had little work performed on exploring the miRNA biology. Primarily, Yi et al. (2013) have sequenced small RNA libraries prepared from shoot samples and identified 43 known miRNAs, 172 novel miRNAs, and 2 mirtron precursor candidates. Among these, 8 were validated and, interestingly, the study found 69 miRNAs to be conserved between foxtail millet and sorghum (Yi et al. 2013). Khan et al. (2014) reanalyzed the publicly available genome sequence of foxtail millet to predict the miRNAs. The study identified 355 mature miRNAs, which were used for target prediction, construction of the physical map, ontology annotation, and validation. Expression profiling of these miRNAs in response to different abiotic stresses suggested their involvement in stress-responsive machinery (Khan et al. 2014). Furthermore, an miRNA database (http://59.163.192.91/FmMiRNADb/index.html) has also been constructed to supply the identified miRNAs to researchers. Recently, Yadav et al. (2016) constructed four small RNA libraries from two contrasting cultivars (tolerant cv. IC403579, sensitive cv. IC480117) of foxtail millet in control and dehydration stress conditions. High-throughput sequencing and analysis revealed the presence of 191 miRNAs, which includes 55 known miRNAs (representing 22 families) and 136 novel candidate miRNAs (representing 48 families). A total of 18 known miRNAs belonging to 15 families and 33 novel miRNAs belonging to 29 families were found to be differentially expressed in response to dehydration stress. In tolerant cv. IC403579, 34 miRNAs were differentially expressed, and the majority of these (32) were found to be up-regulated in response to stress. Interestingly, 22 out of 32 differentially expressed miRNAs were down-regulated after stress treatment in sensitive cultivars (Yadav et al. 2016). Functional characterization such as over-expression of these identified miRNAs and their targets would provide more insight into the differential behavior of foxtail millet cultivars during dehydration stress and help in elucidating the role of miRNAs in regulating dehydration stress response.

11.10 Conclusions

miRNA-mediated post-transcriptional gene regulation is one of the mechanisms conferring stress tolerance to plants, and it involves either cleavage or translation inhibition of corresponding target genes. High-throughput sequencing and miRNA microarray methods have reported many miRNAs that show stress-responsive expression patterns. Induction of miRNAs during stress can down-regulate their target mRNAs, which may be a negative regulator of drought response. On the other hand, down-regulated miRNAs lead to the accumulation of their target mRNAs which might contribute positively to stress adaptation. Extensive investigations have been carried out in many plant species to identify stress-associated miRNAs; however, no few studies were performed to delineate the stress responsive role of miRNAs in foxtail millet (*Setaria italica*), a naturally abiotic stress tolerant model crop. In this context, extensive molecular investigations are requisite to unravel the miRNA-mediated stress regulatory network.

Acknowledgements Studies on millet genomics in Dr. Manoj Prasad's laboratory are supported by Science and Engineering Research Board (SERB), Department of Science and Technology (DST), Govt. of India [Grant No. EMR/2015/000464], by Department of Biotechnology, Govt. of India [Grant No. BT/HRD/NBA/37/01/2014], and by Core Grant of National Institute of Plant Genome Research (NIPGR), New Delhi, India. Hari-Gowthem G acknowledges Department of Biotechnology, Govt. of India for the research fellowship.

References

Achard P, Herr A, Baulcombe DC, Harberd NP (2004) Modulation of floral development by a gibberellin-regulated microRNA. Development 131:3357–3365

Agarwal S, Grover A (2006) Molecular biology. Biotechnology and genomics of flooding-associated low O_2 stress response in plants. Crit Rev Plant Sci 25:1–21

Allen E, Xie Z, Gustafson AM, Carrington JC (2005) MicroRNA-directed phasing during trans acting siRNA biogenesis in plants. Cell 121:207–221

Arenas-Huertero C, Pérez B, Rabanal F, Blanco-Melo D, De la Rosa C, Estrada-Navarrete G, Sanchez F, Covarrubias A, Reyes J (2009) Conserved and novel miRNAs in the legume *Phaseolus vulgaris* in response to stress. Plant Mol Biol 70:385–401

Aukerman MJ, Sakai H (2003) Regulation of flowering time and floral organ identity by a microRNA and its APETALA2-like target genes. Plant Cell 15:2730–2741

Aung K, Lin SI, Wu CC, Huang YT, Su CL, Chiou TJ (2006) pho2, a phosphate over accumulator, is caused by a nonsense mutation in a microRNA399 target gene. Plant Physiol 141:1000–1011

Bailey-Serres J, Voesenek LA (2008) Flooding stress: acclimations and genetic diversity. Annu Rev Plant Biol 59:313–339

Baker CC, Sieber P, Wellmer F, Meyerowitz EM (2005) The early extra petals1 mutant uncovers a role for microRNA miR164c in regulating petal number in *Arabidopsis*. Curr Biol 15:303–315

Bari R, Datt Pant B, Stitt M, Scheible W-R (2006) PHO2 MicroRNA399 and PHR1 define a phosphate-signaling pathway in plants. Plant Physiol 141:988–999

Bartel D (2004) MicroRNAs: genomics, biogenesis, mechanism, and function. Cell 116:281–297

Baumberger N, Baulcombe DC (2005) *Arabidopsis* ARGONAUTE1 is an RNA Slicer that selectively recruits microRNAs and short interfering RNAs. Proc Nat Acad Sci U S A 102:11928–11933

Ben CS, Liu R, Chinnusamy V, Kwon Y, Park JH, Kim SY, Zhu JK, Yang SW, Lee BH (2013) STA1, an *Arabidopsis* pre-mRNA processing factor 6 homolog, is a new player involved in miRNA biogenesis. Nucleic Acids Res 41:1984–1997

Bennetzen JL, Schmutz J, Wang H, Percifield R, Hawkins J, Pontaroli AC, Estep M, Feng L, Vaughn JN, Grimwood J, Jenkins J, Barry K, Lindquist E, Hellsten U, Deshpande S, Wang X, Wu X, Mitros T, Triplett J, Yang X, Ye CY, Mauro-Herrera M, Wang L, Li P, Sharma M, Sharma R, Ronald PC, Panaud O, Kellogg EA, Brutnell TP, Doust AN, Tuskan GA, Rokhsar D, Devos KM (2012) Reference genome sequence of the model plant Setaria. Nat Biotechnol 30:555–561

Bologna NG, Voinnet O (2014) The diversity, biogenesis, and activities of endogenous silencing small RNAs in Arabidopsis. Annu Rev Plant Biol 65:473–503

Bologna NG, Mateos JL, Bresso EG, Palatnik JF (2009) A loop-to-base processing mechanism underlies the biogenesis of plant microRNAs miR319 and miR159. EMBO J 28:3646–3656

Bonnet E, Wuyts J, Rouze P, Van de PY (2004) Detection of 91 potential conserved plant microRNAs in Arabidopsis thaliana and *Oryza sativa* identifies important target genes. Proc Nat Acad Sci U S A 101:11511–11516

Chen X (2004) A microRNA as a translational repressor of APETALA2 in *Arabidopsis* flower development. Science 303:2022–2025

Chen L, Ren Y, Zhang Y, Xu J, Sun F, Zhang Z, Wang Y (2012) Genome-wide identification and expression analysis of heat-responsive and novel microRNAs in *Populus tomentosa.* Gene 504:160–165

Chiou T, Aung K, Lin S, Wu C, Chiang S, Su C (2006) Regulation of phosphate homeostasis by microRNA in *Arabidopsis.* Plant Cell 18:412–421

Chuck G, Cigan AM, Saeteurn K, Hake S (2007) The heterochronic maize mutant Corngrass1 results from over-expression of a tandem microRNA. Nat Genet 39:544–549

Cooper JE (2007) Early interactions between legumes and rhizobia: disclosing complexity in a molecular dialogue. J Appl Microbiol 103:1355–1365

Cuperus JT, Montgomery TA, Fahlgren N, Burke RT, Townsend T, Sullivan CM, Carrington JC (2010) Identification of MIR390a precursor processing-defective mutants in *Arabidopsis* by direct genome sequencing. Proc Nat Acad Sci U S A 107:466–471

Ding D, Zhang L, Wang H, Liu Z, Zhang Z, Zheng Y (2009) Differential expression of miRNAs in response to salt stress in maize roots. Ann Bot 103:29–38

Earley KW, Poethig RS (2011) Binding of the cyclophilin 40 ortholog SQUINT to Hsp90 protein is required for SQUINT function in *Arabidopsis.* J Biol Chem 286:38184–38189

Ellendorff U, Fradin EF, de Jonge R, Thomma BP (2009) RNA silencing is required for *Arabidopsis* defence against *Verticillium* wilt disease. J Exp Bot 60:591–602

Emery JF, Floyd SK, Alvarez J, Eshed Y, Hawker NP, Izhaki A, Baum SF, Bowman JL (2003) Radial patterning of *Arabidopsis* shoots by class III HD-ZIP and KANADI genes. Curr Biol 13:1768–1774

Franco-Zorrilla JM, Valli A, Todesco M, Mateos I, Puga MI, Rubio-Somoza I, Leyva A, Weigel D, Garcia JA, Paz-Ares J (2007) Target mimicry provides a new mechanism for regulation of microRNA activity. Nat Genet 39:1033–1037

Fu C, Sunkar R, Zhou C et al (2012) Over-expression of miR156 in switchgrass (*Panicum virgatum* L.) results in various morphological alterations and leads to improved biomass production. Plant Biotechnol J 10:443–452

Fujii H, Chiou T-J, Lin S-I, Aung K, Zhu J-K (2005) A miRNA involved in phosphate-starvation response in *Arabidopsis.* Curr Biol 15:2038–2043

Gandikota M, Birkenbihl RP, Höhmann S, Cardon G, Saedler H, Huijser P (2007) The miRNA156/157 recognition element in the 3′ UTR of the Arabidopsis SBP box gene SPL3 prevents early flowering by translational inhibition in seedlings. Plant J 49:683–693

Gao P, Bai X, Yang L, Lv D, Li Y, Cai H, Ji W, Guo D, Zhu Y (2010) Over-expression of osa-MIR396c decreases salt and alkali stress tolerance. Planta 231:991–1001

Gao P, Bai X, Yang L, Lv D, Pan X, Li Y, Cai H, Ji W, Chen Q, Zhu Y (2011) osa-MIR393: a salinity- and alkaline stress-related microRNA gene. Mol Biol Rep 38:237–242

Guo HS, Xie Q, Fei JF, Chua NH (2005) microRNA directs mRNA cleavage of the transcription factor NAC1 to down regulate auxin signals for *Arabidopsis* lateral root development. Plant Cell 17:1376–1386

Han MH, Goud S, Song L, Fedoroff N (2004) The *Arabidopsis* double-stranded RNA-binding protein HYL1 plays a role in microRNA-mediated gene regulation. Proc Nat Acad Sci U S A 101:1093–1098

Hirsch J, Lefort V, Vankersschaver M, Boualem A, Lucas A, Thermes C, d'Aubenton-Carafa Y, Crespi M (2006) Characterization of 43 non-protein-coding mRNA genes in Arabidopsis, including the MIR162a-derived transcripts. Plant Physiol 140:1192–1204

Hsieh LC, Lin SI, Shih AC, Chen JW, Lin WY, Tseng CY, Li WH, Chiou TJ (2009) Uncovering small RNA-mediated responses to phosphate deficiency in *Arabidopsis* by deep sequencing. Plant Physiol 151:2120–2132

Iki T, Yoshikawa M, Nishikiori M, Jaudal MC, Matsumoto-Yokoyama E, Mitsuhara I, Meshi T, Ishikawa M (2010) In vitro assembly of plant RNA-induced silencing complexes facilitated by molecular chaperone HSP90. Mol Cell 39:282–291

Iki T, Yoshikawa M, Meshi T, Ishikawa M (2012) Cyclophilin 40 facilitates HSP90-mediated RISC assembly in plants. EMBO J 31:267–278

Jia F, Rock CD (2013) MIR846 and MIR842 comprise a cistronic MIRNA pair that is regulated by abscisic acid by alternative splicing in roots of *Arabidopsis.* Plant Mol Biol 81:447–460

Jones-Rhoades MW, Bartel DP (2004) Computational identification of plant microRNAs and their targets, including a stress-induced miRNA. Mol Cell 14:787–799

Juarez MT, Kui JS, Thomas J, Heller BA, Timmermans MC (2004) microRNA-mediated repression of rolled leaf1 specifies maize leaf polarity. Nature 428:84–88

Kantar M, Lucas S, Budak H (2010) miRNA expression patterns of *Triticum dicocoides* in response to shock drought stress. Planta 233:471–484

Khraiwesh B, Zhu JK, Zhu K (2012) Role of miRNAs and siRNAs in biotic and abiotic stress responses of plants. Biochim Biophys Acta 1819:137–148

Khan Y, Yadav A, Suresh BV, Muthamilarasan M, Yadav CB, Prasad M (2014) Comprehensive genome-wide identification and expression profiling of foxtail millet [*Setaria italica* (L.)] miRNAs in response to abiotic stress and development of miRNA database. Plant Cell Tiss Organ Cult 118:279–292

Kim W, Benhamed M, Servet C, Latrasse D, Zhang W, Delarue M, Zhou DX (2009) Histone acetyl transferase GCN5 interferes with the miRNA pathway in *Arabidopsis*. Cell Res 19:899–909

Kim J, Lee H, Jung H, Maruyama K, Suzuki N, Kang H (2010a) Over-expression of microRNA395c or 395e affects differently the seed germination of *Arabidopsis thaliana* under stress conditions. Planta 232:1447–1454

Kim JY, Kwak KJ, Jung HJ, Lee HJ, Kang H (2010b) MicroRNA402 affects seed germination of *Arabidopsis thaliana* under stress conditions via targeting DEMETER-LIKE Protein3 mRNA. Plant Cell Physiol 51:1079–1083

Kim YJ, Zheng B, Yu Y, Won SY, Mo B, Chen X (2011) The role of mediator in small and long noncoding RNA production in *Arabidopsis thaliana*. EMBO J 30:814–822

Koornneef M, Leon-Kloosterziel KM, Schwartz SH, Zeevaart JAD (1998) The genetic and molecular dissection of abscisic acid biosynthesis and signal transduction in Arabidopsis. Plant Physiol Biochem 36:83–89

Kulcheski FR, de Oliveira LFV, Molina LG, Almerão MP, Rodrigues FA, Marcolino J, Barbosa JF, Stolf-Moreira R, Nepomuceno AL, Marcelino-Guimarães FC, Abdelnoor RV, Nascimento LC, Carazzolle MF, Pereira GAG, Margis R (2011) Identification of novel soybean microRNAs involved in abiotic and biotic stresses. BMC Genom 12:307

Kutter C, Schob H, Stadler M, Meins F Jr, Si-Ammour A (2007) MicroRNA-mediated regulation of stomatal development in *Arabidopsis*. Plant Cell 19:2417–2429

Lagos-Quintana M, Rauhut R, Lendeckel W, Tuschl T (2001) Identification of novel genes coding for small expressed RNAs. Science 294:853–858

Lau NC, Lim LP, Weinstein EG, Bartel DP (2001) An abundant class of tiny RNAs with probable regulatory roles in *Caenorhabditis elegans*. Science 294:858–862

Laufs P, Peaucelle A, Morin H, Traas J (2004) microRNA regulation of the CUC genes is required for boundary size control in *Arabidopsis* meristems. Development 131:4311–4322

Lauter N, Kampani A, Carlson S, Goebel M, Moose SP (2005) microRNA172 down-regulates glossy15 to promote vegetative phase change in maize. Proc Nat Acad Sci U S A 102:9412–9417

Lee RC, Ambros V (2001) An extensive class of small RNAs in *Caenorhabditis elegans*. Science 294:862–864

Lee RC, Feinbaum RL, Ambros V (1993) The *C. elegans* heterochronic gene lin-4 encodes small RNAs with antisense complementarity to lin-14. Cell 75:843–854

Li J, Yang Z, Yu B, Liu J, Chen X (2005) Methylation protects miRNAs and siRNAs from a 3′ end uridylation activity in *Arabidopsis*. Curr Biol 15:1501–1507

Li H, Deng Y, Wu T, Subramanian S, Yu O (2010a) Mis-expression of miR482, miR1512 and miR1515 increases soybean nodulation. Plant Physiol 153:1759–1770

Li T, Li H, Zhang YX, Liu JY (2010b) Identification and analysis of seven H_2O_2 responsive miRNAs and 32 new miRNAs in the seedlings of rice (*Oryza sativa* (L.) ssp. indica). Nucleic Acids Res 39:2821–2833

Li B, Duan H, Li J, Deng X, Yin W, Xia X (2013) Global identification of miRNAs and targets in *Populus euphratica* under salt stress. Plant Mol Biol 81:525–539

Liang G, Yang F, Yu D (2010) MicroRNA395 mediates regulation of sulfate accumulation and allocation in *Arabidopsis thaliana*. Plant J 62:1046–1057

Lin J-S, Lin C-C, Lin H-H, Chen Y-C, Jeng S-T (2012) MicroR828 regulates lignin and H_2O_2 accumulation in sweet potato on wounding. New Phytol 196:427–440

Liu PP, Montgomery TA, Fahlgren N, Kasschau KD, Nonogaki H, Carrington JC (2007) Repression of AUXIN RESPONSE FACTOR10 by microRNA160 is critical for seed germination and post-germination stages. Plant J 52:133–146

Liu HH, Tian X, Li YJ, Wu CA, Zheng CC (2008) Microarray based analysis of stress-responsive microRNAs in *Arabidopsis thaliana*. RNA 14:836–843

Lobbes D, Rallapalli G, Schmidt DD, Martin C, Clarke J (2006) SERRATE: a new player on the plant microRNA scene. EMBO Rep 7:1052–1058

Lu C, Fedoroff N (2000) A mutation in the *Arabidopsis* HYL1 gene encoding a dsRNA binding protein affects responses to abscisic acid, auxin, and cytokinin. Plant Cell 12:2351–2366

Lu SF, Sun YH, Shi R, Clark C, Li LG, Chiang VL (2005) Novel and mechanical stress-responsive microRNAs in *Populus trichocarpa* that are absent from *Arabidopsis*. Plant Cell 17:2186–2203

Lu S, Sun YH, Chiang VL (2008) Stress-responsive microRNAs in *Populus*. Plant J 55:131–151

Lv DK, Bai X, Li Y, Ding XD, Ge Y, Cai H, Ji W, Wu N, Zhu YM (2010) Profiling of cold-stress-responsive miRNAs in rice by microarrays. Gene 459:39–47

Machida S, Chen HY, Adam Yuan Y (2011) Molecular insights into miRNA processing by *Arabidopsis thaliana* SERRATE. Nucleic Acids Res 39:7828–7836

Mallory AC, Dugas DV, Bartel DP, Bartel B (2004a) microRNA regulation of NAC-domain targets is required for proper formation and separation of adjacent embryonic, vegetative, and floral organs. Curr Biol 14:1035–1046

Mallory AC, Reinhart BJ, Jones-Rhoades MW, Tang G, Zamore PD et al (2004b) microRNA control of PHABULOSA in leaf development: importance of pairing to the microRNA 5′ region. EMBO J 23:3356–3364

Mallory AC, Bartel DP, Bartel B (2005) MicroRNA-directed regulation of *Arabidopsis* AUXIN RESPONSE FACTOR17 is essential for proper development and modulates expression of early auxin response genes. Plant Cell 17:1360–1375

Manavella PA, Hagmann J, Ott F, Laubinger S, Franz M, Macek B, Weigel D (2012a) Fast-forward genetics identifies plant CPL phosphatases as regulators of miRNA processing factor HYL1. Cell 151:859–870

Manavella PA, Koenig D, Weigel D (2012b) Plant secondary siRNA production determined by microRNA-duplex structure. Proc Nat Acad Sci U S A 109:2461–2466

McConnell JR, Emery J, Eshed Y, Bao N, Bowman J, Barton MK (2001) Role of PHABULOSA and PHAVOLUTA in determining radial patterning in shoots. Nature 411:709–713

Mi S, Cai T, Hu Y, Chen Y, Hodges E, Ni F, Wu L, Li S, Zhou H, Long C, Chen S, Hannon GJ, Qi Y (2008) Sorting of small RNAs into Arabidopsis argonaute complexes is directed by the 5′ terminal nucleotide. Cell 133:116–127

Millar AA, Gubler F (2005) The *Arabidopsis* GAMYB-like genes, MYB33 andMYB65, are microRNA-regulated genes that redundantly facilitate anther development. Plant Cell 17:705–721

Mittler R, Vanderauwera S, Gollery M, Van Breusegem F (2004) Reactive oxygen gene network of plants. Trends Plant Sci 9:490–498

Moldovan D, Spriggs A, Yang J, Pogson BJ, Dennis ES, Wilson IW (2009) Hypoxia-responsive microRNAs and trans-acting small interfering RNAs in *Arabidopsis*. J Exp Bot 61:165–177

Montgomery TA, Howell MD, Cuperus JT, Li D, Hansen JE, Alexander AL, Chapman EJ, Fahlgren N, Allen E, Carrington JC (2008) Specificity of argonaute7-miR390 interaction and dual functionality in TAS3 trans-acting siRNA formation. Cell 133:128–141

Morozova O, Marra MA (2008) Applications of next-generation sequencing technologies in functional genomics. Genomics 92:255–264

Muthamilarasan M, Prasad M (2015) Advances in *Setaria* genomics for genetic improvement of cereals and bioenergy grasses. Theor Appl Genet 128:1–14

Muthamilarasan M, Prasad M (2017) Genetic determinants of drought stress tolerance in *Setaria*. In: Doust A, Diao X (eds) Genetics and genomics of Setaria. Springer, pp 267–289

Navarro L, Dunoyer P, Jay F, Arnold B, Dharmasiri N, Estelle M, Voinnet O, Jones JD (2006) A plant miRNA contributes to antibacterial resistance by repressing auxin signaling. Science 312:436–439

Navarro L, Jay F, Nomura K, He SY, Vionnet O (2008) Suppression of the microRNAs pathway by bacterial effector proteins. Science 321:964–967

Nikovics K, Blein T, Peaucelle A, Ishida T, Morin H, Aida M, Laufs P (2006) The balance between the MIR164A and CUC2 genes controls leaf margin serration in *Arabidopsis*. Plant Cell 18:2929–2945

Nozawa M, Miura S, Nei M (2012) Origins and evolution of microRNA genes in plant species. Genome Biol Evol 4:230–239

Park W, Li J, Song R, Messing J, Chen X (2002) CARPEL FACTORY, a Dicer homolog, and HEN1, a novel protein, act in microRNA metabolism in *Arabidopsis thaliana*. Curr Biol 12:1484–1495

Pasquinelli AE, Reinhart BJ, Slack F, Martindale MQ, Kuroda MI, Maller B, Hayward DC, Ball EE, Degnan B, Müller P, Spring J, Srinivasan A, Fishman M, Finnerty J, Corbo J, Levine M, Leahy P, Davidson E, Ruvkun G (2000) Conservation across animal phylogeny of the sequence and temporal regulation of the 21 nucleotide let-7 heterochronic regulatory RNA. Nature 408:86–89

Perez-Quintero AL, Quintero A, Urrego O, Vanegas P, Lopez C (2012) Bioinformatic identification of Cassava miRNAs differentially expressed in response to infection by *Xanthomonas axonopodis* pv manihotis. BMC Plant Biol 12:29

Pieczynski M, Marczewski W, Hennig J, Dolata J, Bielewicz D, Piontek P, Wyrzykowska A, Krusiewicz D, Strzelczyk-Zyta D, Konopka-Postupolska D, Krzeslowska M, Jarmolowski A, Szweykowska-Kulinska Z (2013) Down-regulation of CBP80 gene expression as a strategy to engineer a drought-tolerant potato. Plant Biotechnol J 11:459–469

Piriyapongsa J, Jordan IK (2008) Dual coding of siRNAs and miRNAs by plant transposable elements. RNA 14:814–821

Qi Y, Denli AM, Hannon GJ (2005) Biochemical specialization within *Arabidopsis* RNA silencing pathways. Mol Cell 19:421–428

Ramachandran V, Chen X (2008) Degradation of microRNAs by a family of exoribonucleases in Arabidopsis. Science 321:1490–1492

Reinhart BJ, Slack FJ, Basson M, Bettinger JC, Pasquinelli AE, Rougvie AE, Horvitz HR, Ruvkun G (2000) The 21 nucleotide let-7 RNA regulates developmental timing in *Caenorhabditis elegans*. Nature 403:901–906

Reinhart BJ, Weinstein EG, Rhoades MW, Bartel B, Bartel DP (2002) MicroRNAs in plants. Genes Dev 16:1616–1626

Ren G, Xie M, Dou Y, Zhang S, Zhang C, Yu B (2012) Regulation of miRNA abundance by RNA binding protein TOUGH in *Arabidopsis*. Proc Nat Acad Sci U S A 109:12817–12821

Ren G, Xie M, Zhang S, Vinovskis C, Chen X, Yu B (2014) Methylation protects microRNAs from an AGO1-associated activity that uridylates 5′ RNA fragments generated by AGO1 cleavage. Proc Nat Acad Sci U S A 111:6365–6370

Reyes JL, Chua NH (2007) ABA induction of miR159 controls transcript levels of two MYB factors during Arabidopsis seed germination. Plant J 49:592–606

Rhoades M, Reinhart B, Lim L, Burge C, Bartel B (2002) prediction of plant microRNA targets. Cell 110:513–520

Schommer C, Palatnik JF, Aggarwal P, Chételat A, Cubas P, Farmer EE, Nath U, Weigel D (2008) Control of jasmonate biosynthesis and senescence by miR319 targets. PLoS Biol 6:e230

Schwab R, Palatnik JF, Riester M, Schommer C, Schmid M, Weigel D (2005) Specific effects of microRNAs on the plant transcriptome. Dev Cell 8:517–527

Sieber P, Wellmer F, Gheyselinck J, Riechmann JL, Meyerowitz EM (2007) Redundancy and specialization among plant microRNAs: role of theMIR164 family in developmental robustness. Development 134:1051–1060

Slack FJ, Basson M, Liu Z, Ambros V, Horvitz HR, Ruvkun G (2000) The lin-41 RBCC gene acts in the *C. elegans* heterochronic pathway between the let-7 regulatory RNA and the LIN-29 transcription factor. Mol Cell 5:659–669

Smith MR, Willmann MR, Wu G, Berardini TZ, Moller B, Weijers D, Poethig RS (2009) Cyclophilin 40 is required for microRNA activity in *Arabidopsis*. Proc Nat Acad Sci U S A 106:5424–5429

Stief A, Altmann S, Hoffmann K, Pant BD, Scheible WR, Bäurle I (2014) *Arabidopsis* miR156 regulates tolerance to recurring environmental stress through SPL transcription factors. Plant Cell 26:1792–1807

Subramanian S, Fu Y, Sunkar R, Barbazuk WB, Zhu JK, Yu O (2008) Novel and nodulation-regulated microRNAs in soybean roots. BMC Genom 9:160

Sunkar R, Zhu JK (2004) Novel and stress-regulated microRNAs and other small RNAs from *Arabidopsis*. Plant Cell 16:2001–2019

Sunkar R, Kapoor A, Zhu J (2006) Posttranscriptional induction of two Cu/Zn superoxide dismutase genes in *Arabidopsis* is mediated by downregulation of miR398 and important for oxidative stress tolerance. Plant Cell 18:2051–2065

Trindade I, Capitão C, Dalmay T, Fevereiro M, Santos D (2010) miR398 and miR408 are up-regulated in response to water deficit in *Medicago truncatula*. Planta 231:705–716

Vandenabeele S, Van-Der-Kelen K, Dat J, Gadjev I, Boonefaes T, Morsa S, Rottiers P, Slooten L, Van-Montagu M, Zabeau M, Inze D, Van-Breusegem F (2003) A comprehensive analysis of hydrogen peroxide-induced gene expression in tobacco. Proc Nat Acad Sci U S A 100:16113–16118

Vazquez F, Gasciolli V, Crete P, Vaucheret H (2004) The nuclear dsRNA binding protein HYL1 is required for microRNA accumulation and plant development, but not posttranscriptional transgene silencing. Curr Biol 14:346–351

Wang F, Perry SE (2013) Identification of direct targets of FUSCA3, a key regulator of arabidopsis seed development. Plant Physiol 161:1251–1264

Wang JW, Wang LJ, Mao YB, Cai WJ, Xue HW, Chen XY (2005) Control of root cap formation by microRNA-targeted auxin response factors in *Arabidopsis*. Plant Cell 17:2204–2216

Wang JW, Schwab R, Czech B, Mica E, Weigel D (2008) Dual effects of miR156-targeted SPL genes and CYP78A5/KLUH on plastochron length and organ size in *Arabidopsis thaliana*. Plant Cell 20:1231–1243

Wang Y, Sun F, Cao H, Peng H, Ni Z, Sun Q, Yao Y (2012) TamiR159 directed wheat TaGAMYB cleavage and its involvement in anther development and heat response. PLoS ONE 7:e48445

Wightman B, Ha I, Ruvkun G (1993) Posttranscriptional regulation of the heterochronic gene lin-14 by lin-4mediates temporal pattern formation in *C. elegans*. Cell 75:855–862

Wilkinson S, Davies WJ (2002) ABA-based chemical signaling: the coordination of responses to stress in plants. Plant, Cell Environ 25:195–210

Wu G, Poethig RS (2006) Temporal regulation of shoot development in *Arabidopsis thaliana* by miR156 its target SPL3. Development 133:3539–3547

Wu L, Zhou H, Zhang Q, Zhang J, Ni F, Liu C, Qi Y (2010) DNA methylation mediated by a microRNA pathway. Mol Cell 38:465–475

Wu F, Shu J, Jin W (2014) Identification and validation of miRNAs associated with the resistance of maize (*Zea mays* L.) to *Exserohilum turcicum*. PLoS ONE 9: e87251

Xie Z, Kasschau KD, Carrington JC (2003) Negative feedback regulation of Dicer-Like1 in *Arabidopsis* by microRNA-guided mRNA degradation. Curr Biol 13:784–789

Xie Z, Allen E, Wilken A, Carrington JC (2005) DICER-LIKE 4 functions in trans-acting small interfering RNA biogenesis and vegetative phase change in *Arabidopsis thaliana*. Proc Nat Acad Sci U S A 102:12984–12989

Xin M, Wang Y, Yao Y, Xie C, Peng H, Ni Z, Sun Q (2010) Diverse set of microRNAs are responsive to powdery mildew infection and heat stress in wheat (*Triticum aestivum* L.). BMC Plant Biol 10:123–129

Xin M, Wang Y, Yao Y, Song N, Hu Z, Qin D, Xie C, Peng H, Ni Z, Sun Q (2011) Identification and characterization of wheat long non-protein coding RNAs responsive to powdery mildew infection and heat stress by using microarray analysis and SBS sequencing. BMC Plant Biol 11:61

Yadav A, Khan Y, Prasad M (2016) Dehydration-responsive miRNAs in foxtail millet: genome-wide identification, characterization and expression profiling. Planta 243:749–766

Yamaguchi-Shinozaki K, Shinozaki K (2006) Transcriptional regulatory networks in cellular responses and tolerance to dehydration and cold stresses. Annu Rev Plant Biol 57:781–803

Yan K, Liu P, Wu CA, Yang GD, Xu R, Guo QH, Huang JG, Zheng CC (2012) Stress-induced alternative splicing provides a mechanism for the regulation of microRNA processing in *Arabidopsis thaliana*. Mol Cell 48:521–531

Yang L, Liu ZQ, Lu F, Dong AW, Huang H (2006) SERRATE is a novel nuclear regulator in primary microRNA processing in *Arabidopsis*. Plant J 47:841–850

Yang S, Tang F, Gao M, Krishnan HB, Zhu H (2010) R gene-controlled host specificity in the legume-rhizobia symbiosis. Proc Nat Acad Sci U S A 107:18735–18740

Yang GD, Yan K, Wu BJ, Wang YH, Gao YX, Zheng CC (2012) Genome-wide analysis of intronic microRNAs in rice and Arabidopsis. J Genet 91:313–324

Yang J, Liu X, Xu B, Zhao N, Yang X, Zhang M (2013) Identification of miRNAs and their targets using high-throughput sequencing and degradome analysis in cytoplasmic male-sterile and its maintainer fertile lines of *Brassica juncea*. BMC Genom 14:9

Yant L, Mathieu J, Dinh TT, Ott F, Lanz C, Wollmann H, Chen X, Schmid M (2010) Orchestration of the floral transition and floral development in Arabidopsis by the bifunctional transcription factor APETALA2. Plant Cell 22:2156–2170

Yi F, Xie S, Liu Y, Qi X, Yu J (2013) Genome-wide characterization of microRNA in foxtail millet (*Setaria italica*). BMC Plant Biol 13:212

Yu B, Yang Z, Li J, Minakhina S, Yang M, Padgett RW, Steward R, Chen X (2005) Methylation as a crucial step in plant microRNA biogenesis. Science 307:932–935

Yumul RE, Kim YJ, Liu X, Wang R, Ding J, Xiao L, Chen X (2013) POWERDRESS and diversified expression of the MIR172 gene family bolster the floral stem cell network. PLoS Genet 9:e1003218

Zeng Q-Y, Yang C-Y, Ma Q-B, Li X-P, Dong WW, Nian H (2012) Identification of wild soybean miRNAs and their target genes responsive to aluminum stress. BMC Plant Biol 12:182

Zhang B (2015) MicroRNA: a new target for improving plant tolerance to abiotic stress. J Exp Bot 66:1749–1761

Zhang J, Xu Y, Huan Q, Chong K (2009) Deep sequencing of Brachypodium small RNAs at the global genome level identifies microRNAs involved in cold stress response. BMC Genomics 10:449

Zhang JF, Yuan LJ, Shao Y, Du W, Yan DW, Lu YT (2008) The disturbance of small RNA pathways enhanced abscisic acid response and multiple stress responses in *Arabidopsis*. Plant, Cell Environ 31:562–574

Zhang W, Gao S, Zhou X, Chellappan P (2011) Bacterial responsive miRNAs regulate plant innate immunity by modulating plant hormone networks. Plant Mol Biol 75:93–105

Zhang X, Xia J, Lii YE, Barrera-Figueroa BE, Zhou X, Gao S, Lu L, Niu D, Chen Z, Leung C, Wong T, Zhang H, Guo J, Li Y, Liu R, Liang W, Zhu JK, Zhang W, Jin H (2012) Genome-wide analysis of plant nat-siRNAs reveals insights into their distribution, biogenesis and function. Genome Biol 13:R20

Zhang LW, Song JB, Shu XX, Zhang Y, Yang ZM (2013a) miR395 is involved in detoxification of cadmium in *Brassica napus*. J Hazard Mater 15 (250–251):204–211

Zhang S, Xie M, Ren G, Yu B (2013b) CDC5, a DNA binding protein, positively regulates posttranscriptional processing and/or transcription of primary microRNA transcripts. Proc Nat Acad Sci U S A 110:17588–17593

Zhao M, Ding H, Zhu J-K, Zhang F, Li W-X (2011) Involvement of miR169 in the nitrogen-starvation responses in *Arabidopsis*. New Phytol 190:906–915

Zhou L, Liu Y, Liu Z, Kong D, Duan M, Luo L (2010) Genome-wide identification and analysis of drought-responsive microRNAs in *Oryza sativa*. J Exper Bot 61:4157–4168

Zhou M, Li D, Li Z, Hu Q, Yang C, Zhu L, Luo H (2013) Constitutive expression of a miR319 gene alters plant development and enhances salt and drought tolerance in transgenic creeping bentgrass. Plant Physiol 161:1375–1391

Zhu H, Zhou Y, Castillo-Gonzalez C, Lu A, Ge C, Zhao YT, Duan L, Li Z, Axtell MJ, Wang XJ, Zhang X (2013) Bidirectional processing of pri-miRNAs with branched terminal loops by *Arabidopsis* Dicer-like1. Nat Struct Mol Biol 20:1106–1115

MIX
Papier aus verantwortungsvollen Quellen
Paper from responsible sources
FSC® C105338

If you have any concerns about our products,
you can contact us on
ProductSafety@springernature.com

In case Publisher is established outside the EU,
the EU authorized representative is:
Springer Nature Customer Service Center GmbH
Europaplatz 3, 69115 Heidelberg, Germany

Printed by Libri Plureos GmbH
in Hamburg, Germany